T0093749

CHEAPER, FASTER, BETTER

CHEAPER, FASTER, BETTER

HOW WE'LL WIN
THE CLIMATE WAR

TOM STEYER

Spiegel
and Grau

Spiegel & Grau, New York
www.spiegelandgrau.com

Interior design by Meighan Cavanaugh

Library of Congress Cataloging-in-Publication Data Available Upon Request

ISBN 978-1954118645 (hardcover)
ISBN 978-1954118652 (eBook)

Printed in the United States of America

Cheaper, Faster, Better is FSC certified. It is printed on chlorine-free paper
made with 30 percent post-consumer waste. It uses only vegetable-based ink.
The binders boards are 100 percent recycled.

First Edition
10 9 8 7 6 5 4 3 2 1

To my amazing kids—Sam, Gus, Evi, and Henry.
And to my longest and fiercest partner, the talented Kat Taylor.

CONTENTS

CHEAPER, FASTER, BETTER

INTRODUCTION

I n 1981, years before I founded an investment fund that grew to have billions of dollars under management, and decades before I walked away from that fund to focus full-time on fighting climate change, I worked out of the Alaska state headquarters of Exxon.

I was twenty-four years old at the time. Business school was starting in the fall, and my job that summer was to help the state figure out how to spend its windfall tax revenue. In Alaska, "windfall tax revenue" means "oil money." That's why our offices were in Exxon's building.

At the time, I didn't see any contradiction between the work I was doing and the kind of person I wanted to be. Hardly anyone thought seriously about climate change back then. I certainly didn't. I liked the work. I *loved* Alaska. I had never thought of myself as an environmentalist or conservationist, but I had always spent as much time as possible outdoors, and the Last Frontier was, by far, the most beautiful place I'd ever seen.

One landscape in particular made a powerful impression on me: a snow-filled valley stretched between two mountains outside Anchorage. Something about that spot—at once sparse and magnificent—stuck in my mind, so much so that twenty-five years later, in 2006, I brought my family up there. I couldn't wait to show Kat and our four kids the beautiful landscape, to see their faces as they took in the vast glacial expanse just as I had a quarter century before.

But it was gone.

I'd read about climate change before that trip, but I'd always thought of it as something that might happen in theory, or on the margins, or in the very distant future, or somewhere far away from me. In an instant, I realized I'd been wrong. As an investor, I'd spent my career looking for patterns, assessing risks, responding to disruption, and looking around corners. I liked to think I was good at my job. And staring at this empty space, where a massive glacier had taken eons to form and less than thirty years to vanish, three things became suddenly clear:

First, climate change was real—and happening much faster than most of us imagined at the time.

Second, climate change would affect us all: economies, governments, businesses, societies. *This will cause famines*, I thought. *This will cause wars.*

The third thing I realized was perhaps the most important, although it was less a realization than a deep, immediate conviction: We can and must solve it.

I've always been an optimist. Life's more fun that way. But my faith that humanity can overcome the climate crisis—and that America can and must lead the way—isn't just a reflection of my personality. My parents belonged to the Greatest Generation, the one that fought and won World War II. When I was a kid, when two grown-ups met, they didn't ask each other, "What do you do for a living?" They asked, "What did you do in the war?" My father, for instance, was a successful corporate lawyer, but the accomplishment he was most proud of—by far—was serving as a prosecutor in the Nuremberg trials while in the Navy.

The America my parents taught me about was truly the home of the brave. In the face of the greatest threat our planet had ever faced, we didn't lose our nerve or self-confidence. We did our part, as a country and as individuals. We transformed ourselves, on the battlefield and the home front, to meet the moment and lead the world to victory.

Maybe that's why, as I learned more about the science of climate change and began to understand how quickly the planet was changing, I never felt depressed or overwhelmed. I did, however, start to believe that there was something wrong with the way I was living my life. I'd built and run a highly successful firm. I could pick up the phone and call my governor or senator about the issues I cared about. I'd even become, much to my surprise, a billionaire. But I wasn't satisfied. This wasn't the life I'd imagined. As much pride as I took in building a successful business, I wanted to be part of a great American success story, the way my parents' generation had been.

On January 1, 2013, I walked away from the business I'd founded and built over decades, and I devoted myself to fighting climate change.

Since then, my partner of nearly forty years, Kat Taylor, and I started a regenerative ranch dedicated to proving that you can raise cattle and have a negative carbon footprint. I co-founded Galvanize Climate Solutions, a new business that bets on companies that we believe can help us save the planet. I donated more than a quarter of a billion dollars to Democratic campaigns and causes, more than any other individual, making nearly all my Republican friends mad at me. Then, when I felt that neither party was focusing enough on climate, I entered the Democratic primary for president myself, making nearly all my Democratic friends mad at me.

I've now spent more than fifteen years immersed in the science, politics, finance, and technology behind the fight to protect our planet, and ourselves, from climate change. But as important as those areas are, none of them is the reason I left my business behind to become a full-time climate activist. Instead, what drove me to change my life is the question I kept asking myself, one that kept me up at night in those years after I returned from my family trip to Alaska. *Protecting humanity from climate change is the fight of our lifetime. Am I doing my part?*

That question is what this book is about.

I'm not saying that you should drop everything and devote yourself full-time to climate—at least not necessarily. But I am saying you should think about it. If you want to lead an interesting, rewarding, and fulfilling life, the kind of life that makes a difference, there's nowhere better to do it than as part of the climate movement. And there have never been more ways to join that movement than there are right now.

Though you don't have to completely upend your life to do your part, there's a good chance you'll have to make big changes. To be a responsible person in today's world, you need to understand the basics of what's happening to our planet, and how it affects us all. Most important, you have to make sure your actions reflect that understanding, not just at the margins, but by incorporating climate into every big decision you make. In other words, whether you work with your hands or in front of a screen, whether you're in business, tech, finance, government, politics, the arts, the sciences, nonprofits, health care, or, really, any other sector, the journey is the same. Whether you redefine your life or just alter it, if you're not there already, you need to become a climate person.

What that means in practice is different for everyone. So while I can't tell you exactly what path you should take, I'd like to share some things I've learned that helped me as I became a climate person—and I hope they can help you, too. Many of these ideas aren't useful exclusively when it comes to climate; they can be applied to investing, business, and any number of other areas. But climate is what matters most right now, and nothing else comes close. "What are you doing to fight climate change?" is the "What did you do in the war?" of our time.

With that in mind, this book is divided into three sections. The first section is about *why*. If you aren't yet convinced you should become a climate person—maybe you think you can't make a big enough difference, or feel that it's too late in your life and career to make a change, or that climate change is overwhelming—I hope this section will help convince you. (If you don't need convincing but are looking to convince others, I hope this section will help you do that, too.)

The second section is about *how*. Once you've made the decision to incorporate climate into your life, I'll discuss some of the ways you can change your approach to the world, and to those around you, so that you can have the biggest impact.

The third and final section is *where*. I don't mean that in a geographical sense. Instead, I describe in detail four areas where we, as a movement and a planet, need to make rapid progress to solve the climate crisis. No matter who you are, or where your talents lie, you should be able to feel at home in at least one of these areas.

One thing I won't spend much time on is debating whether climate change is real. That debate is settled. In fact, by this point, nearly all of us have experienced climate change firsthand. Look at what happened just last year, in the United States alone. If you're from Phoenix, you experienced thirty-one straight days of temperatures over 110 degrees Fahrenheit. If you're a New Yorker, you watched the sky go dark as your city was blanketed by smoke from wildfires hundreds of miles away. If you're from South Florida, you stepped into an ocean that was literally as hot as a Jacuzzi. According to a poll conducted by the Associated Press, 90 percent of Americans say they experienced extreme weather due to climate change last summer. No wonder that even the oil companies admit global temperatures are rising and that extreme weather is becoming more common.

But most people, even most climate-conscious people, still don't realize that three trends have completely reset the trajectory of our planet, not over the last century but in the last few years.

Two of these trends are pretty depressing, and one is extremely encouraging. I'll start with the bad news. The science—by which I mean the facts on the ground, as measured and understood by the most objective experts on climate—is much worse than we thought it was just a few years ago. For decades, we've known the basics: burning fossil fuels (mainly oil, gas, and coal) releases pollution (mainly carbon and methane gas) into our atmosphere more quickly than forests and oceans can absorb them. All this gas acts like a blanket for the earth, trapping heat and raising temperatures. As that happens, all kinds of weather patterns—patterns on which human beings have relied since the dawn of civilization—are disrupted.

What's new is just how quickly this is occurring. The oceans are warming faster than scientists projected, especially at the poles. Greenland and Antarctica are losing much more ice to melting than we predicted, which will mean a greater rise in sea level. In 2022, the level of Antarctic sea ice was 9 percent below its historical average. In 2023, that number jumped to 17 percent. We don't know what the exact number will be for 2024; all we know is that sooner or later, we're going to lose enough ice to cause oceans to rise by several feet, and that hundreds of millions of people live within several feet of what we currently think of as sea level.

Climate isn't like flipping a switch. When people say, "We have X number of years to solve this or we're all doomed," or, "We have to limit warming to such-and-such a level or humanity's going extinct," they're oversimplifying things. The way I think about it is that while climate change is already certain to affect our lives, if we remain on the track we're on, it will soon *define* our lives. In such a scenario, global warming won't end human existence, but it will upend human society as we know it—not just in some places, but everywhere, the United States very much included.

Which leads me to the second depressing trend: the oil and gas companies don't care. Throughout this book, I'm going to use a few

interchangeable terms—"oil and gas," "fossil fuels," "the fossil fuel industry," and so on. What I'm referring to, however, are not just the people
who run companies that extract and burn oil, gas, and coal. Instead,
I'm talking about an entire ecosystem: businesspeople, banks, insurers,
politicians, consultants, tax lawyers, media figures, think tanks, and so
many others who benefit from fossil fuels and the ridiculous amount
of money they generate while ignoring the devastating costs to society.

For a while, it seemed that the oil and gas companies might transform themselves—mostly because they thought that elected officials and
grassroots movements would force them to. But in the last few years their
calculus has changed. Yes, their ads talk about doing the right thing for
the planet. But if you look at the investments they're making, and what
they're saying to their shareholders, they actually expect us to use just as
much oil and gas in 2050 as we do today, or perhaps even more. (Just to
be clear: if we're using as much oil and gas in 2050 as we use right now,
the individuals in the fossil fuel industry will be really rich—but the
planet will be really, really screwed.)

Today, oil and gas companies are behaving the way cigarette companies behaved in the 1990s. They know their products are hurting people,
but as long as the money keeps flowing, they'll keep doing what they're
doing. There are two big differences, though. First, oil and gas receives
an estimated $7 *trillion* in government subsidies each year. Second, oil
and gas doesn't just threaten individuals who burn it or who are exposed
to its fumes. If we allow climate change to go unchecked, it will affect
every person on this planet.

But despite the doom and gloom we hear so often these days, I don't
think we'll let that happen. Because the other trend shaping our planet
right now is this: as bad as the facts on the ground have gotten, and as
determined as the fossil fuel industry is to resist any kind of meaningful change, clean-energy tech is much, much further ahead than most
people realize.

You've probably started to see some of this yourself. The cost of wind and solar have plummeted. Electric vehicles are on the roads. But these are only a few highly visible examples of a much larger phenomenon: clean energy is beating the fossil fuel industry in the marketplace. Just a few years ago, the argument climate people were making was basically about long-term thinking: "Let's pay a little more for climate-friendly stuff now to avoid paying way more in disaster relief later." While that argument was absolutely correct, it asked people to put aside their short-term interests, which is always difficult.

As the world has changed, the argument has changed, too. Today's case for climate action is a lot simpler and more persuasive: "Let's use stuff that works better and more reliably, makes people safer and healthier, and saves everyone money. And let's protect the planet in the process." That's not yet true for everything—airplanes, for example, are hard to run using anything but fossil fuels. But the same revolution we saw in EVs and solar is happening everywhere (including aviation), and it's happening fast. Not every promising new technology or early-stage company will succeed. But some will. And they'll completely change the way we think about everything from energy to economics to international affairs.

So here's where we stand right now. Climate change is already harming our planet, and if we stay on our current path, climate catastrophe will devastate it. The oil and gas companies are fighting hard to maintain the status quo. But despite their money, political clout, seemingly endless government subsidies, and decades of goodwill and respect in elite circles, we've already turned a corner.

When I first started thinking about writing this book about two years ago, I would have said that we were in the climate equivalent of December 8, 1941, the day after Pearl Harbor was attacked and Americans realized that they couldn't ignore the war, whether they wanted to or not. Today I'd say we're in the equivalent of 1943. There are tough battles ahead. There are going to be plenty of difficult days, and heartbreaking news to go with them. But it's clearer than ever that we can win. And not just that. If everyone does their part, we're *going* to win. I have no doubt about it.

What does winning look like? It's a crucial question. Like many others, I often describe the fight against climate change as a fight to protect the planet, to save the planet, to ensure that Earth remains habitable, or to avert a global crisis. But while all these things are true, at least in a broad sense, we've reached a point where we need to be more specific. After all, there's no day on which we'll be able to declare the planet saved or protected. At the same time, looking at the extreme weather upending lives around the world, it's clear that a global climate crisis can no longer be completely prevented. It's already begun. If left unchecked, that crisis will grow exponentially more dire. There's a lot at stake if we don't minimize the damage, but we'll never be able to entirely turn back the clock.

There is, however, a more specific definition for winning on climate. If we can get to a point where the amount of global greenhouse gas pollution emitted into the atmosphere is balanced out by the amount absorbed—what the climate movement often calls "net zero"—it will be the climate equivalent of V-E Day. We'll still have a lot of work to do, and the pollution already in the atmosphere will continue to affect us. But for the first time in modern history, we'll no longer be contributing to climate change. And once that happens, we can truly begin to rebuild.

That's the fight I'm hoping you'll join—the fight to stabilize the planet.

Prevailing in this fight is the only hope we have for a sustainable and healthy future. And it's a fight we're already starting to win. Net zero used to feel like a pipe dream. Today, it feels like an inevitability. It's also become clear that moving toward that goal won't just improve our planet's future—it will improve our lives right away. The journey to net zero means breathing cleaner air, producing more energy at lower cost, creating new products that run better and more reliably for less money, and developing new technologies that have the potential to improve just about every aspect of society. In the same way that technologies developed to win World War II helped lay the groundwork for everything from modern computing and agriculture to surfboards and Slinkys, the fight to stabilize our planet will launch us into a brand new, better-than-ever scientific and economic age.

At the same time, every year that goes by until we reach that inevitable, net-zero moment means more damage and suffering, over both the short and long term. If we delay too long, it's even possible that rising temperatures will melt the tundra and release vast amounts of methane into the atmosphere, causing climate change to spiral out of our control. If that were to happen, a worst-case scenario for humanity would become all but certain.

The decision to become a climate person, then, has never been more meaningful or urgent. But there's something else I want to add from my personal experience: being a climate person is fun! Some people imagine that to be on the right side of this issue, you have to don a hairshirt, stop eating hamburgers, and be a downer at parties. Not true at all! I can say with complete conviction that getting involved in fighting climate change won't ruin your life. It will *improve* it.

Someone once asked me how much money I left on the table in order to turn my attention to climate. I did some quick math. The answer? Between three and ten billion dollars—and probably a lot closer to ten than three.

But I wouldn't trade these last fifteen years for anything. They've been the most interesting, rewarding, educational, and downright enjoyable of my professional life. And for the record, while I made less money by focusing on climate, most people who go into climate today won't have to. In fact, if you care about getting rich, being part of the revolution taking place in energy is a pretty good way to do it.

Trying to prevent a global catastrophe is worth doing in and of itself. But the climate movement is no longer about conservation, at least not in the traditional sense of keeping the world the way it is right now. It's about making the world better than it's ever been before. It's about a future where energy is cheaper, cleaner, and more reliable for everyone. We're not trying to get rid of the products, inventions, and devices that define modern life. We're trying to upgrade them so that they cost less, work better, spew less pollution into the air and keep you and your family healthier. In the process, we can free the world from the outsized influence of autocracies like Russia, Iran, and Saudi Arabia, who get to hold the planet hostage just because they have a bunch of dead dinosaurs in the ground.

I'm not naïve. I've spent years studying the science. But I absolutely remain an optimist. If we can make it through the next decade, we won't just prevent the most catastrophic consequences of climate change. We'll be living in the single best moment in human history.

For the last decade, I've defined success, in everything I do, in terms of impact on our planet. That goes for this book too. I'm thrilled you're reading it, and I hope lots of other people will read it too. But this book is just a means to an end. The world is changing, and chances are it's changing a lot faster than you realize. If this book can help convince you to be a part of a community working to change the way we think, and even more so change the way we act, so that we can build a better world, I'll have succeeded.

We can stabilize the planet, prevent unimaginable human suffering, lead the world into a new American century, and build a future better than anything we've ever dreamed of. I really believe that.

And I want you to believe it, too.

CHAPTER I

DO THE OBVIOUS THING

Over the last few decades, and especially over the last few years, the fight against climate change has become part of the culture wars. That didn't happen by accident. The oil and gas industry worked hard to turn what is at heart an engineering problem into one of the most emotionally charged, polarizing issues of our time.

So let's start our discussion of climate by taking out the word "climate" entirely. Instead, I want to tell you about two competing, generic industries—we'll call them A and B.

Industry A is a hotbed of new ideas and is experiencing explosive growth. Industry B is doing basically the same thing it's done for the past century. An estimated $1.7 trillion was invested in Industry A last year, 70 percent more than the amount invested in Industry B. The prices of Industry A's marquee products have fallen by nearly 100 percent in the past few decades. The prices of Industry B's products are volatile—so much so that we take for granted the way its price spikes regularly strain households and rattle governments around the world—but the average cost hasn't fallen in at least fifty years. Industry B relies on trillions of dollars in annual government subsidies to remain competitive; despite being subsidized far less, many of Industry A's products are already substantially cheaper than their Industry-B equivalents.

Also, Industry A will prevent enormous amounts of human suffering, while Industry B is causing enormous amounts of human suffering—leading people and governments around the world to turn against Industry B.

Here's the big question: knowing all this, which industry would you rather invest in? Which one would you rather work for? Which one would you want to support—whether with your purchasing power or your vote?

As you've probably gathered, Industry A is clean energy, and B is fossil fuels. While stabilizing our planet will be difficult, deciding to join the climate movement shouldn't be. All it requires, given everything going on in the world today, is a willingness to make an extremely obvious choice.

When I think about the importance of the obvious, I remember a conversation I had with my friend Warren Hellman. In 1986, I left one of the most coveted jobs on Wall Street to strike out on my own. I moved to San Francisco, hardly a city synonymous with global finance, especially back then. Not long afterward, I opened my investment fund, Farallon Capital. It was just me and an assistant. We didn't even have our own office. But Warren ran Hellman & Friedman, a small merchant bank in town, and with an introduction and some persuading from Matt Barger—one of Warren's most trusted partners and one of my closest friends—he agreed to let us share his space on Market Street.

An upside of all this was that I got to spend a lot of time with Warren, who, in addition to being a great guy, was a very good investor. How good? Well, about a decade before we moved into his office, he placed a big bet on a small company called Apple Computer.

One day, as we were working out of the office we shared, I asked Warren what he had seen in Apple that others hadn't. I expected him to credit a stroke of genius, or a flash of insight, or a sophisticated computer model. But that's not what he said. Instead, laughing, he told me this:

"Because it was one of the most obvious things of all time!"

The point he was making was not just about Apple. Investing is about predicting the future. That's hard. But predicting the future isn't difficult for the reasons you might think. Many people assume that successful investors are daredevils—taking long-shot chances that turn large profits through a combination of incredible luck and skill. But in my experience, you don't get credit for degree of difficulty. Taking wild risks and hoping they pay off isn't investing, it's gambling. The real work of a good investor is in identifying the sure things that, for whatever reason, others overlook.

What's true for investing money is also true for investing time—and right now, as we enter this new phase of the fight against climate change, the question of how you should invest your time has never been more crucial. Just as it did for those in my parents' generation, "doing your part" will look different for everyone. But it all begins with the same choice: the decision to spend a significant portion of your time trying to help stabilize our planet and prevent the enormous human tragedy climate change is poised to inflict.

No matter who you are, what you do, or what stage of life you're in, devoting serious time to protecting our planet from climate change is an investment I'm 100 percent sure will be worth it. Why am I so confident? Because at the end of the day, when you become a climate person, you're doing the same thing that Warren did when he bet big on Apple: you're doing one of the most obvious things of all time.

Of course, the "obvious thing" isn't obvious to everybody. What I also learned as an investor is that uncovering obviousness is a skill—one we can all learn. When you examine the world in the right way, and you ask the right kinds of questions, you'll find that you start to see the future in a way that others often can't, and you can then make decisions accordingly.

This requires different approaches in different situations. But as I learned early on in my career, it can be as simple as finding the relevant numbers and adding them up.

There's a good chance you're too young to remember Braniff Airways, and if you do remember them, you probably haven't thought about them for decades. But when I was growing up, they were a major player. Founded in 1927 as a kind of flying timeshare for a handful of rich Oklahomans, Braniff became a major carrier, with 115 aircraft and 81 destinations around the world at its peak.

Then it fell apart. In the late 1970s the airline industry was deregulated, increasing competition. At the same time, for a variety of reasons including the Iranian Revolution, the price of oil—and with it, jet fuel—skyrocketed. After a few rapid-succession changes in leadership, Braniff ceased air operations in 1982. By the end of the year, the company was finished, turned over to its creditors.

While all this was happening, I was a young associate in the risk arbitrage division of Goldman Sachs. "Risk arbitrage" sounds fancy, but it basically works like this: you make an investment based on something you think will happen that's related to an event you already *know* will happen. In this case, we knew the Braniff reorganization was definitely happening—the company was being split into parts, which would then be sold off. Creditors in the company had been given certificates entitling them to a portion of the proceeds from that sale. They could either sell those certificates immediately, just as you'd sell a government bond, or they could wait and redeem their certificates after the fire sale was over.

I'm sure you've heard the expression "missing the forest for the trees." This was the opposite. A lot of creditors started off with a sweeping conclusion (*This airline is bankrupt. Get me out of here!*) and sold their certificates as quickly as they could. But my boss, the legendary Bob Rubin—who later went on to run Goldman and then become Bill

Clinton's treasury secretary—asked me to look into the matter more closely to figure out what the certificates were really worth.

Let's say, just to make the numbers easy, that each certificate was worth 1 percent of the company and was selling for one dollar. If we predicted the fire sale was going to make more than 100 dollars total, the certificates were cheap and we should buy them; if we predicted they would make less than 100 dollars, the certificates were overpriced and we shouldn't. I got to work. Which is to say, I dug into mountains of paper files (this was before the internet) and tried to figure out what everything Braniff owned was worth. How much did a used 727 sell for? What price could you get for the rights to a gate at an airport in Dallas, or Honolulu, or LA?

It wasn't an easy process. It took a lot of time, which is probably why most people didn't do it. But after a while, I'd figured out all the numbers and added them up. I almost did a double take. This stuff was worth a lot! The company wasn't viable, but the assets the company owned had value—way more value than the price of the certificates suggested. It was as though stacks of ten one-dollar bills were selling for five dollars apiece.

We started buying. At one point, I think we owned 35 percent of a defunct airline. And when the liquidation occurred, and the parts were sold off, we made a lot of money on the deal. I was happy about that, but I wasn't surprised. It felt like betting on two plus two being four.

If you want to figure out the obvious thing, sometimes it's as simple as starting with the trees rather than the forest: understanding what the relevant numbers are, putting them together, and seeing what they say.

And what the numbers say on climate couldn't be clearer.

I'm not a scientist, any more than I was an airline industry executive when I dug into Braniff. But you don't need to be a scientist to put together the pieces of climate change. Even back in 2006, when some people were still claiming climate change wasn't real, this wasn't hard to figure out.

Here's the short version of the different things we know—the trees that make up the forest.

For more than a century, burning fossil fuels allowed us to do incredible things. We discovered how to turn oil and gas into fuel for cars and airplanes and stoves; into chemicals and plastics to simplify our daily lives; into heat for our homes or to manufacture new products; and into electricity to power all our modern appliances and devices. Best of all, compared to the forms of energy that came before them, fossil fuels were cheap. They were like magic. A miracle drug. But like many miracle drugs, they had a side effect: the greenhouse gas they released into the air caused global temperatures to rise.

It's tempting to imagine the planet as a giant room, with carbon and methane turning up the thermostat by a few degrees. Such a scenario, while uncomfortable, wouldn't be devastating. Unfortunately, that's not how the earth works. A better comparison would be to the human body. If your internal temperature is 98.6, you feel fine. If that number rises by just one degree Fahrenheit, you start to feel unwell. If your temperature goes up by just a few degrees more, it's a matter of life and death. Systems you count on to keep you alive and well—systems that usually run so seamlessly that you take them for granted—begin to break down, with each failure causing a cascade of new failures.

Right now, the thermometer is rising across the globe. As we know—both because we've measured it and because, at this point, we've seen it with our own eyes—the global warming caused by burning fossil fuels is devastating the planet and threatening all the progress fossil fuels once helped us achieve. While knowing what the exact weather somewhere will be ten years from now is impossible, when it comes to the big picture on climate, it's pretty easy to predict what the future will look like, in broad strokes, if we don't act. Here are some of the not-so-tough questions facing humanity when it comes to climate:

- The Gulf Stream—which pumps warm water north through the Atlantic—is the reason that most of Europe has temperate weather, even though the continent is very far north. Melting ice from the poles is already slowing that process down. If the Gulf Stream stops flowing entirely, and Europe is plunged into an ice age even as the rest of the planet grows dangerously warmer, how will that turn out?

- Natural disasters are devastating to economies. There's not just the cost of rebuilding (usually borne by taxpayers); there's the cost of small businesses shuttered, the skyrocketing price of insurance for homeowners and employers in a disaster's wake (or the inability to purchase insurance at any price), the loss of income of people who work outdoors who have to cut back on their hours due to rising temperatures, not to mention the human suffering and deaths that accompany these catastrophes.

  During the 2000s, the United States experienced an average of about seven disasters per year that cost $1 billion or more to recover from. During the 2010s, that number jumped to thirteen billion-dollar disasters per year, and it's been even higher during the 2020s. If this trend doesn't just continue but accelerates, will that be good or bad for the economy?

- Worse weather isn't just costly. It's harmful to our health and well-being. The number of climate-related deaths is already growing. So, in many parts of the country, is the number of days during which we're forced to stay inside because of heat, torrential downpours, and wildfire smoke that just a decade ago would have been anomalous but today have become routine. If the number of heat waves or dangerous air-quality days each summer increases, and they last longer, is that going to help or hurt our quality of life?

- A few years ago, persistent drought in Central America was a big driver of immigration to the United States. People couldn't feed their families, so they made the agonizing decision to flee their homes for somewhere safer to live. Think about the way a few hundred thousand refugees helped fan the flames of the reactionary politics encouraged by people like Donald Trump. Now imagine that instead of hundreds of thousands of desperate people, we're talking about hundreds of millions—and consider that every part of the world is experiencing an identical refugee crisis simultaneously. Based on how we've done so far dealing with much smaller crises, is our geopolitical system going to handle that well?

These are not difficult questions to answer.

But it was also clear back in 2006—and it's even clearer now—that we can prevent the worst effects of climate change before it's too late. In the mid-1800s we developed oil and gas to replace existing forms of energy. Today we need to develop sources of energy as good as or better than fossil fuels, but without the dangerous side effects. And we need to do it in a way that allows developing countries to leapfrog oil and gas, so that as they expand their economies, they can go straight to cleaner sources of energy without having to use fossil fuels as a stepping stone.

What's interesting is that today, even in a politically divided country, most people understand this. A survey conducted by Yale researchers in October 2023 found that 71 percent of Americans believe global warming will harm future generations, 73 percent would like to see regulation of carbon dioxide as a pollutant, and 69 percent believe we should transition the US economy to 100 percent clean energy by 2050. The point is that when you dig into the evidence on climate, which these days includes reading the news or looking out the window, most people don't have a hard time figuring out what's going on or where we're headed.

Climate change is the most dangerous global threat facing humanity right now. And that's not a controversial statement, or at least it shouldn't be. It's like saying two plus two is four.

But though adding up the numbers is probably the easiest way to identify the obvious choice, it's not always possible. Sometimes, instead of looking at numbers, we find ourselves looking at stories—hypothetical scenarios that lead to different futures—and deciding which one to believe.

That's what happened to me in 1985, when Great Britain's ninety-nine-year lease on Hong Kong approached its final decade. The return of Hong Kong to China was significant for all sorts of geopolitical reasons: as the end of a major piece of Britain's colonial legacy; as a potential threat to the island's then-stable democracy (which unfortunately has backslid enormously in recent years); as a signal of China's rising influence throughout the Pacific.

The pending expiration of the lease was also a huge deal in the financial markets, because under British rule, Hong Kong had become a major Asian financial hub, with a stock market second in size only to Tokyo's. What would happen to the stocks of companies listed on that market the day after Hong Kong was, quite literally, taken over by communists?

Chinese officials tried over and over again to reassure the public that trading would continue as usual after the handoff. But the markets weren't reassured. Stock prices plummeted. At that point, a typical price-to-earnings ratio for a stock on the Tokyo exchange was somewhere between twelve-to-one and twenty-five-to-one. On the Hong Kong exchange, it fell to between two-to-one and three-to-one. (To put it differently, a stock that would have been worth at least $100 a share on the Tokyo exchange was worth no more than $25, just because it was listed in Hong Kong.) People worried that Hong Kong might shut down the day the PRC took over.

I found myself asking the same question I asked when I was dealing with Braniff Airways certificates: what were these assets really worth? There was, however, a key difference. I had no numbers to add up. I couldn't know what the market value of a stock listed on the Hong Kong exchange was after Hong Kong was returned to China because Hong Kong hadn't been returned to China yet.

What I had to do instead was choose between competing stories. In one story, Chinese officials would leave the Hong Kong stock exchange unaltered, at least in the short term. In another story, communists would immediately restrict, or even shut down, market trading. If the first story was true, the stocks were worth far more than they were selling for. We should buy. If the second story was true, the stocks had plummeted in price for good reason and might fall further. We should avoid them like the plague and sell any we already owned.

We bought. Unlike with Braniff Airways, I wasn't 99.9 percent sure my decision would turn out to be correct. But I was still extremely confident, and here's why: the buy story was much, much simpler than the sell story. The Chinese government had promised to leave the markets untouched, and they had no reason to lie—Hong Kong was being handed over to the PRC either way. What's more, China was desperate to grow its economy and had strong incentives not to jeopardize relationships with major companies and potential trading partners.

Just about all the Asia experts I talked to shared this view. They worried, with good reason, about the long-term prospects for Hong Kong's democracy and economy. But they couldn't find a plausible explanation for why China would want to blow up the Hong Kong stock exchange on day one.

The sell story wasn't impossible. But it was much more complicated. It would involve Chinese officials making unnecessary pronouncements, then deciding to go back on their word even though it was against their economic interest to do so. You could see how such a thing might happen—but it would take a chain reaction of unlikely events.

When Hong Kong was eventually handed over to the PRC, both stories were put to the test. And what happened next? The Chinese government allowed the market to continue functioning, just as they said it would. Stock prices surged back up as the threat receded. People who bet on the more complicated story did poorly, and those who bet on the simpler story did really well.

Today, when it comes to climate, each of us has to choose between two competing stories about the future. I want to compare those stories, and do so with just one question:

Which climate story is simplest?

Let's start by looking at the story told by oil and gas when they try to convince their investors—not to mention the public—that we can keep burning fossil fuels at our current rates. Basically, it relies on some combination of things happening. One, the earth will self-regulate its climate indefinitely—even though there's no evidence or scientific basis for thinking that will happen. Two, we'll soon be able to make carbon-capture technology so cheap that we can afford to remove all the greenhouse gas we emit from the atmosphere—even though most projections from those in the carbon-capture industry say that we won't be able to counteract more than 20 percent of our current emissions by 2050. Three, if the first two predictions don't work, we'll be able to geo-engineer our planet using unproven technologies that have vast, irreversible consequences—without any serious negative effects.

I'm not saying all that is literally impossible. We're talking about the future, where, by definition, nothing is 100 percent certain. But the fossil fuel industry's story requires an elaborate combination of good luck, sheer coincidence, and outrageous levels of inaccuracy on the part of every serious scientist and researcher working on climate today.

To call the fossil-fuel story complicated doesn't even begin to describe it.

Now let's look at the climate-people story.

To do that, I want to go back to 2010, when an out-of-state oil company got a measure onto the California ballot that would undo nearly all the climate measures our state legislature had passed. This was in the middle of the Great Recession—unemployment was more than 9 percent—and the fossil fuel lobby was spending a ton of money on ads arguing that we just couldn't afford pro-climate policies. At the time, the conventional wisdom in California was that you didn't fight the oil companies, because they had too much money and always won. But the brilliant Chris Lehane, a longtime political strategist and friend, called me up and said, "You're always talking about climate, here's your chance to do something." So I decided to lead the fight against the ballot measure.

First, though, I needed a Republican partner, and Chris thought of the perfect guy—my good friend, a lifelong Republican, and Ronald Reagan's former secretary of state, the late George Shultz. George was well-respected, especially among conservatives. After leaving the Reagan Administration, he'd become a fellow at the highly conservative Hoover Institution, joined the board of Chevron, and did all kinds of other things successful Republicans from that era did. No one would ever accuse him of being a bleeding heart or a socialist.

But George was a data-driven guy, and I knew he could see the numbers on climate just like anyone else. And because of his background, he could make his case to the business groups and the chambers of commerce and they'd listen.

George agreed to come on board, and what he'd say was this: "You have fire insurance because it's unlikely your house is going to burn down, but if it burns down it's a devastating loss. If your business had a 20 percent chance of bankruptcy, you'd have to deal with it, even though there's still an 80 percent chance everything would be fine. Well, what if there's a 20 percent chance that the entire planet might destabilize? Don't you think it's worth taking out an insurance policy against that?"

George knew how to speak the language of conservative business leaders. He got many prominent Republicans, including then-governor Arnold Schwarzenegger, to endorse our side—and he got others to stay neutral. Our goal wasn't just to defeat the oil companies at the ballot box. George always said to me, "We don't just need to beat these guys. We need to smash them." And we did. Our winning margin was more than 23 percent.

As it happens, I think George greatly understated things. If we act too slowly on climate, the odds of global catastrophe aren't 20 percent, they're probably more like 99 percent. He also left out the part of the story that involves the benefits of clean energy—if new technologies could not just prevent catastrophe but could provide cheap, clean, reliable power for everyone, wouldn't it be worth developing those technologies as fast as we possibly can? But even in George's telling, which downplayed both the costs of doing nothing and the benefits of positive action, the climate-person story is incredibly simple: we're facing a remarkably dangerous threat, and we should do something about it.

Do I sometimes wonder if the fossil-fuel story about the future is true—if the earth, for example, has some secret self-regulating mechanism unknown to science? Sure I do. But it seems tremendously unlikely. In fact, it's so unlikely that the oil and gas companies' own scientists have repeatedly said that they don't think it's possible. And here's the thing: if it turns out that against all odds, the fossil fuel industry's complicated explanation for why we can keep emitting carbon at our current rates is accurate, then we have nothing to worry about. But if we bet on their story and it *isn't* true, we won't be able to correct our mistake. By the time we find out, it will be too late.

Choosing among different methods for fighting climate change—policy approaches, technologies, and financing models, for example—might be complicated. But deciding whether to go all-in on fighting climate change or not just requires us to decide whose story to believe. And that couldn't be more obvious.

With both Braniff Airways and the Hong Kong stock exchange, there's a similarity that goes beyond the chance to buy something at a discount. In each case, it was clear that the status quo was about to change dramatically. Braniff was going to be liquidated. Hong Kong was going to revert from British to Chinese control. In most cases, however, there's less advance warning that a major change is coming. And that often leads people to make one of the biggest mistakes when trying to predict the future: they rely on the entirely unreasonable assumption that the status quo will persist forever. But human history shows us that the opposite is true. In fact, one of the best quotes about identifying the obvious comes from a *Wall Street Journal* op-ed written by the conservative economist Herbert Stein:

"If something cannot go on forever, it will stop."

It's easy to think that the status quo will go on forever because of the nature of large-scale change. People imagine change happening linearly, at the same pace every year. Usually it doesn't work like that. Instead, things change slowly, or not at all, for a very long time. For a while, people who bet on the status quo look really smart. Then, the dam breaks, and the rate of change goes from a snail's pace to breakneck speed. The status-quo people don't look so smart anymore.

Take the example of local newspapers. Around the turn of the century, the internet became a brand-new, easily accessible source of news. It soon became obvious to most media-industry observers that there was no longer any reason for people to advertise in the classified pages now that they could do it online. But for several years, nothing really changed. Advertisers kept paying money to appear in the classified section, even though it didn't make sense for them to do so. They clung to their old ways. Newspaper revenue snuck down, but it was nothing like the catastrophe one might have predicted. During that time, if you argued that the threat posed to newspapers by the internet was real, you probably would have been called an alarmist. If you argued that the threat was overstated, you probably would have felt pretty smug.

Then, seemingly overnight, the classified pages didn't just shrink, they basically vanished. Local news went from a slow and steady decline into a tailspin.

This kind of pattern, where a barely noticeable downward incline drops off a full-on cliff, has repeated itself throughout history. Which is probably why history is full of examples of people who refuse to see—or admit they see—the writing on the wall. Unwilling to accept that things will ever change, they cling to the way things are.

Take the American whaling industry. In the mid-1840s, whale oil was a precious commodity, used in lamps to light homes and businesses long before the advent of electric lighting. America was the world leader in whaling. From Nantucket to the West Coast, whaling made some people wildly rich and provided a good living for many more. At one point, it was the fifth-largest industry in the United States. But even at what appeared to be its peak, the industry was dying. New sources of energy, like kerosene and petroleum, were becoming cleaner, safer, cheaper, and more abundant every year. Meanwhile, whales themselves were becoming scarce—fleets had to travel farther to reach them.

As Eric Jay Dolin writes in *Leviathan*, his history of American whaling, many people did the obvious thing: they left the whaling industry for Pennsylvania, the new center of petroleum production. But others, especially those with major financial interests in whaling, became dead-enders. In 1843, the *Nantucket Inquirer* sneered at those who thought "Lard Oil, Chemical Oil, Camphene Oil, and a half dozen other luminous humbugs" would replace whale oil. In 1852, *Whalemen's Shipping List* argued that whale-oil alternatives were dangerous and should be banned by the government.

As late as 1867, Massachusetts senator Charles Sumner publicly lauded "the Whale, whose corporal dimensions fitly represent the space which he occupies in the Fisheries of the world, hardly diminished by petroleum or gas." Like a lot of people with political and economic interests, Sumner became a kind of dead-ender. He was a smart guy, and a brilliant politician. But in the face of ever-growing evidence that the whaling industry was being replaced, he didn't listen. He kept betting on blubber instead.

The *Inquirer*, the *Shipping List*, and Senator Sumner were all ignoring an important truth: things change. The cost of lighting a house with petroleum kept going down, while the cost of lighting it with whale oil kept going up. New innovations made lamps fueled by whale-oil alternatives safer, so that Americans no longer had to choose between affordability and risk. Politicians from whale-oil-producing states like Massachusetts might have been an exception, but lawmakers weren't exactly eager to follow the *Shipping List*'s suggestion of a ban when new technologies were saving people money and creating jobs in the process.

By 1878, the whaler Alexander Starbuck could, despite his industry affiliations, comfortably write of petroleum, "Its dangerous qualities at first greatly checked its general use, but, these removed, it entered into active, relentless competition with whale-oil, and it proved the more powerful of the antagonistic forces." In other words, while whale oil didn't disappear completely, the fight between it and newer forms of energy could now be described, even by whalers, in the past tense.

I'm not trying to be subtle about this, nor am I the first to make this observation: the whaling industry in the mid-1800s sounds a lot like the fossil-fuel industry today. Nearly two centuries ago, a product that had been a seemingly irreplaceable part of modern life was gradually rendered obsolete. The same thing is happening today. Compared to the alternatives, oil and gas are becoming less economically competitive every year—especially in America, where we've already extracted everything that was inexpensive to extract. New technologies are making alternatives to oil and gas more abundant and reliable.

In fact, I'd argue that the oil and gas industry is in a much worse place today than the whaling industry was back then. Why? It's pretty simple. Whaling killed the whales. Oil and gas are killing us.

Yet plenty of people are still betting that the fossil fuel era will go on forever. For many years, I invested in a private equity fund, one where the partners have done an amazing job of generating returns for decades. They're first-class, high-integrity people. But about four years ago, I was surprised to find out that this firm had spent more than $4 billion on a software company that does software for oil and gas drilling in the United States of America.

I was really concerned, for two reasons. First, I thought we'd agreed to avoid investing in fossil fuel companies. Second, and no less important, I worried about the investment from a purely capitalistic, maximize-returns point of view. I said, "What are you guys doing investing in a company that helps drill in the United States? What we drill today, or what we try to drill today, we're going to be producing in ten years. So, you're making a bet that in ten years, all that American oil and gas is still going to be valuable. And on top of that, you're doing software for these companies. So in ten years, when you want to sell your software company, you're going to be doing projections for ten years after *that*. You're betting on oil, in 2041, in the United States, when the cheapest oil in the world is probably going to be in Saudi Arabia, Venezuela, and Russia?"

In other words, for this company to be valuable down the road, global oil production would have to remain at extremely high levels over the next twenty or thirty years. But that's a doomsday scenario for the planet. And more important, at least from the perspective of a monetary return on investment, that's a bet that the status quo will remain essentially unchanged, even though just about everyone agrees it has to change.

That didn't seem like a good bet to me, regardless of what industry it's made in.

While I've been wrong plenty of times as an investor, this was one of those cases where my prediction was correct: the company struggled. But that's not the end of the story. Recently, the company added a new line of business. Today, they're using the technology they developed to help boost fossil fuel production to help boost renewable energy production, and they're doing better. It's just another example of how the smart money is recognizing the huge opportunities that come with the transition to clean energy. They're betting on the future.

I suspect this trend is only going to accelerate in the years ahead. Because the American people (not to mention people around the world) will respond to changes in the natural world that they can see and feel. That's just human nature. We care about our health, our sense of safety, and our children's futures. Already, most Americans want to see a more robust climate response —according to a survey conducted by the nonpartisan Pew Research Center last year, more than two-thirds of American adults say the United States should focus on developing renewable energy over expanding fossil fuel production, and 74 percent support the United States engaging internationally on climate. As natural disasters get worse and wildfire smoke chokes our cities and sea levels rise, wouldn't you expect those numbers to get higher? Do you think we'll sit idly by and let the oil and gas companies wreck our lives?

Of course we won't. If we want to preserve our way of life, burning fossil fuels can't go on forever. And that means it's going to stop. To me, that's the most powerful argument for the idea that saving ourselves from the worst of climate change is both doable and inevitable. I just can't believe that we'd say, "We're going to let our world decay because it's too much of a pain to get off our rear ends and fix it."

But there's a big difference between protecting our world sooner than later. And that's where you—and every other climate person—come in.

The world we're living in is changing, in a way that will render fossil fuels, and all their terrible side effects, obsolete. But if you look at the numbers right now, that change isn't happening fast enough. Powerful forces in business and politics are pushing back to keep the status quo in place for as long as possible. The technology to bring cheaper, cleaner, more reliable energy to everyone is improving rapidly. It's already here in many cases, but not yet in all of them. And the political will to act isn't materializing nearly as fast as we need it to.

We have a choice to make, and we have to make it quickly. I've been a full-time climate activist for more than a decade, but I still face this question every morning: Is there more I can do? Can I devote more of my time, energy, resources to what is clearly the most dangerous threat humanity has ever faced?

Asking yourself if you can and should do more on climate might be a difficult question. But it's not a complicated one. We know what the obvious thing is.

CLIMATE PEOPLE

Tim Heidel and Steve Ashworth

Most people don't realize it, but in America, the clean energy projects that have already been proposed for development could account for 80 percent of the energy our country needs. That's a huge accomplishment—and it means that producing enough renewable energy to get to net-zero emissions is well within our reach.

But generating electricity is just the first step. We also have to get that electricity to people who need it, via our energy grid. That's a problem because our grid is near its maximum capacity. While more renewable projects are being developed every year, many of them are prevented from actually being used to supply power.

The simplest way to increase the grid's capacity is to build more transmission lines. But building new transmission lines is expensive and, generally speaking, the people generating the electricity are responsible for the full cost of connecting new projects to the grid. Building new lines also requires permits that are painstakingly slow to get, and inevitably the lines need to go through land owned by people who don't want power lines in their backyard. "A freakin' nightmare," was how one former advisor to the US Federal Energy Regulatory Commission described the bureaucratic process of modernizing the grid. This, of course, plays right into the fossil fuel industry's hands. Oil and gas loves to point to the difficulty of building new lines to argue that powering our country with renewables is a pipe dream, one that's possible in theory but can never happen in practice.

Enter Tim Heidel and Steve Ashworth, two founders of a Massachusetts-based company called VEIR.

Engineers by trade, Tim and Steve have spent their careers working to advance clean energy, doing everything from conducting research at government laboratories to connecting start-ups with financing to inventing energy-efficient power cables. In 2020, they teamed up to develop a way to increase the capacity of the energy grid—without building new transmission lines.

For a long time, engineers have known it was theoretically possible to use high-temperature superconductors to increase the amount of energy standard power lines can carry by up to 500 percent. However, the refrigeration systems required to cool those superconductors were so heavy that they could only be installed underground.

VEIR invented a much lighter refrigeration system, one that can be installed on transmission lines overhead rather than underground. That means high-temperature superconductors can be used widely on existing lines. It's exactly the type of innovation that my co-founder, Katie Hall, and I launched Galvanize to help jump-start.

Once again, a fossil fuel talking point is being belied by American ingenuity. Expanding the energy grid's capacity fivefold could enable hundreds of clean energy sources to be added without any new digging for transmission lines, putting us on the path to mass electrification. It would also save Americans money on their electric bills, by making clean, inexpensive electricity widely available in large quantities. And with the ability to move electricity efficiently over long distances, VEIR's superconductor technology could soon be used to help unlock the full potential of renewable energy. Imagine solar arrays in the Sahara supplying power across Africa, or wind farms along France's coast providing electricity to Europe.

"There's an urgent need for transmission development all over the world," says CEO Tim Heidel. Every day, he and the team at VEIR are working hard to make it happen—for the sake of the planet and the billions of people living on it.

CHAPTER II

SHARPEN
YOUR BULLSHIT DETECTOR

n December 2022, a dead humpback whale washed ashore on a New
Jersey beach. A few weeks later, another whale was found on a different beach, then another. By summer 2023, eleven dead whales had
been found on the state's beaches—an unusually high number. The story
made local news, then national news, as people began asking what was
killing the whales.

It wasn't long before groups aligned with oil and gas had a scapegoat in
mind: wind energy. New Jersey doesn't have any oil or gas underground,
but it has lots of wind off its coastline, and as part of a transition to clean
energy, the state had authorized the construction of wind turbines ten
to twenty miles from the shoreline. Before you can build an offshore
turbine, you need to map the seafloor using sonar, and an organization
called Protect Our Coast New Jersey told anyone who would listen that
this underwater mapping was behind the whale deaths.

There was no evidence to support this theory, but that didn't seem
to matter. A group of mayors in New Jersey, almost all of whom were
Republicans, wrote a letter to federal officials and the governor opposing

wind farms, saying they were "concerned about the impacts these projects may already be having on our environment." Republican congressmen Jeff Van Drew and Chris Smith, who represent parts of the Jersey Shore, called for a moratorium on turbines. Donald Trump—whose only real previous interest with the Jersey Shore had involved stiffing contractors on his Atlantic City casino projects—added a line about dead whales to his speeches. Tucker Carlson called wind farms "the DDT of our times."

If you took the time to research Carlson's claims, it wasn't hard to see that they were almost certainly false. Scientists at National Oceanic and Atmospheric Administration found that there is no evidence that the types of sonar used in underwater mapping could kill whales. Then there was the physical evidence: not every whale could be autopsied, but those that could be were, and it turned out that they all had been hit by boats. This made sense: whales have been spotted in greater-than-usual numbers off the New Jersey coast over the past several years, and last year, increased shipping resulted in increased boat traffic to and from ports.

Also, while some of the groups that pushed the wind-turbines-killing-whales theory were from the Jersey Shore, when you followed the money you learned that their donors were not. They were backed by a conservative Delaware-based organization called the Caesar Rodney Institute, whose leadership includes David Stevenson, a climate denier and former member of Donald Trump's EPA transition team. The Caesar Rodney Institute, in turn, gets a large portion of its funding from the American Fuel & Petrochemical Manufacturers, a trade association that represents the fossil fuel industry, and the American Energy Alliance, a pro–fossil fuel group started by a former Enron executive. Congressmen Smith and Van Drew, meanwhile, have taken tens of thousands in donations from the oil and gas industry.

That boat collisions were by far the most likely cause of the whale deaths, or that the people arguing against wind energy had major conflicts of interest didn't stop the anti-wind theory from spreading. In fact, opposition to wind energy—based, in large part, on a feigned concern for whales—became a key part of the Republican Party's message going into New Jersey's 2023 statewide elections. Support for wind energy along the Jersey Shore, which had once been strong and bipartisan, fell to just 50 percent. Eventually, citing rising interest rates, the company hired to build the wind farms backed out of the project.

I bring all this up for two reasons. First, it shows that despite their outward confidence, the fossil fuel companies see clean energy as a major threat. They claim that wind energy won't work—that it won't be reliable or cost-effective. But their real concern is that it *will* work. That's why they're spending so much time and money trying to make sure clean energy projects never get up and running.

Second, and perhaps even more important, this anti-wind campaign is a good example of a type of argument that's been employed time and time again by the fossil fuel industry. What made "Sonar mapping kills whales" so insidious is that, as the old saying goes, "You can't prove a negative." There was no good evidence to support the theory, and plenty of good evidence refuting it. But because not every whale could be autopsied, it was impossible to say, with 100 percent confidence, what caused every death. Unlike, say, Trump's provably false claim that wind turbines cause cancer, it was always theoretically possible that some smoking-gun piece of evidence could be uncovered tomorrow, or the next day, or the day after that.

Best of all, from the perspective of the fossil fuel industry, debunking the anti-wind theory and untangling the web of financial ties between oil and gas and supposedly grassroots groups took large amounts of time and effort. Their propaganda campaign had an impact—but even if it hadn't, people in the climate movement would have been forced to switch their focus from helping people understand the benefits of wind energy to rebutting a series of emotionally charged but factually incorrect talking points. For oil and gas companies, it was "heads I win, tails you lose."

There's a word for the fossil fuel industry's go-to tactic: bullshitting.

This isn't the first time bullshitting has been used as the last refuge of a dying industry. Just consider the story of Gene Freidman, a profane, ponytailed New Yorker known as "The Taxi King."

Freidman grew up in a taxi family—his father left the Soviet Union for Queens in the 1970s, became a driver, and ran his own garage—and at the height of his wealth controlled a fleet of hundreds of New York City yellow cabs. Less than a decade ago, if you hailed a cab in the city, there was a pretty good chance it was one of his.

New York City requires every taxi owner to hold a "medallion" to operate a cab. The supply of medallions was severely limited by the city, which made them expensive, and Freidman rented out the many medallions he owned to drivers who couldn't afford to buy them. But that was just the first step. "I'd go to an auction," he once bragged to *Bloomberg*, "I'd run up the price of a medallion, then I'd run to my bankers and say, 'Look how high the medallion's priced! Let me borrow against my portfolio.' And they let me do that."

For a long time, it seemed like a foolproof strategy—the price of a medallion shot from around $160 in today's dollars when they were introduced to more than $1.3 *million* by 2010. With 250 medallions, the Taxi King was at the helm of an empire. As long as prices kept going up, which is to say, as long as speculators thought prices would keep going up, that empire would never stop growing.

Then came Uber.

Like local newspapers and whale oil, this was one of those cases where, at first, the status quo seemed immutable. In fact, for a few years, the price of taxi medallions continued to rise. But by 2013, it started dropping—at first a little, then a lot. Gene Freidman was in trouble.

Today, even though it might not yet seem like it, the fossil fuel companies are in the same position the Taxi King found himself in a little more than a decade ago. They're still making a ton of money. But new tech is threatening their business, growing better and more popular every year.

That's why it's worth paying close attention to how the Taxi King responded when a technological revolution threatened to render his business obsolete. He attacked Uber head-on, calling it "the nastiest, most morally corrupt company ever." He suggested they'd never provide the kind of reliable servicing he could. ("Uber is a brilliant, fascinating technology company. I don't think they're a good taxi company.") Then he pointed to some facts, stripped from context: "I make money. They lose money."

These statements weren't technically lies. It was true that at the time Uber was losing money. And the question of whether a good rideshare company was a good taxi company was mostly subjective. Whether Uber was corrupt and nasty was similarly difficult to prove, and merely debating the issue played into Freidman's hands by ignoring questions about the Taxi King's own business practices. Yet even though Gene Freidman's statements were not provably false, their purpose was to leave the rest of us with a highly misleading impression.

One way you can tell that the oil and gas industry is in trouble is that they're doing today what Gene Freidman did when he went after Uber. They're not lying outright (at least not always). But they're acting with total disregard for the truth, bluffing, weaseling, spinning, saying whatever they can to keep the music playing for as long as possible. They want to convince us that we can keep doing what we're doing, and that everything will turn out fine.

Then there's the other, more personal way that fossil fuel companies are bullshitting you. They want to convince you, either implicitly or explicitly, that becoming a climate person isn't worth the trouble. They want you to believe that devoting a significant portion of your time to protecting the planet is unnecessary, ineffective, extreme, or all of the above. Their profits depend, in large part, on keeping the climate move-ment as small as possible for as long as possible.

There are, of course, two big differences between oil and gas and the Taxi King. First, the taxi industry wasn't always a paragon of virtue, but it wasn't threatening human civilization as we know it. Second, Gene Freidman was one guy. Oil and gas is quite possibly the most politically powerful industry on earth. We're facing what might just be the largest, most well-funded bullshit machine in human history. And the fate of humanity depends on whether we fall for it.

So if you're going to do your part on climate, you need to sharpen your bullshit detector.

I began to develop my own bullshit detection skills at an early age. Many of my earliest memories are of playing cards with my mom, dad, and brothers: Hearts. Poker. Spades. We never played for money; the point was to bluff, to deceive, to put one another on and see what we could get away with. My parents' favorite entertainers, like Groucho Marx and W. C. Fields, were consummate bullshit artists. "There's a sucker born every minute, and two to take him," my father would often warn me as we walked through New York City. Sometimes he'd even point out some of his favorite grifting techniques, like the way W. C. Fields folded a bill in half so the rubes would double-count it. He would have been furious if I'd ever attempted to use that trick on anyone in the real world, but he found it hilarious in movies—and he thought it was important to teach his children how to avoid getting played. My parents were optimists—but they kept their eyes wide open.

Not all Steyers got the message right away. One day, when one of my brothers was home from college, he told me he'd lost 125 bucks playing three-card monte in the park. To a teenager in the 1970s, that was like losing a billion dollars.

I remember asking him, "Did you really think you could win money playing three-card monte? You know how it works."

"No, no, no, no, no," my brother protested. "I watched it for a really long time, and I figured it out, and the guy ahead of me made a lot of money."

I said, "Let me ask you a question: did that guy ahead of you look a lot like the guy behind the table? Did they dress the same? Did they act the same? Did they maybe seem like they knew each other?"

The fossil fuel industry is way more sophisticated than a couple of guys running a three-card monte scam in the park. But when it comes to bullshitting the public, they're often not that different. They'll make what seems, at first, to be a pretty appealing case. When you look a little bit closer, however, you'll notice that everyone making that case seems to know one another—and more important, everyone stands to make a lot of money if the rest of us buy what they're selling.

A good example involves "bridge fuel." You'll often hear fossil fuel executives say something like this: "Coal is dirty, and we need something much cleaner. But we can't switch to renewable energy overnight. So, over the next several years, natural gas can provide a cleaner, temporary alternative as we move away from coal."

For the record, here's the somewhat complicated, technical reason the bridge-fuel argument doesn't work. It's true that gas, while dirtier than renewables, is much cleaner than coal in the lab. But that's because in the lab, gas doesn't leak. In the real world, leaks happen all the time: during transportation along pipelines, from the well, at power plants, and in homes and businesses. According to a recent study by the Environmental Defense Fund, a gas pipeline incident occurs somewhere in the US approximately every forty hours—and leaks are likely to be even more common in developing countries, which have smaller budgets, less oversight, and worse infrastructure.

All these leaks send methane into the atmosphere. From 2010 to 2021, in the US alone, leaks reported to the federal government caused over 26.6 *billion* cubic feet of methane to be released into the atmosphere.

That's even worse for the planet than leaking an equivalent amount of CO_2. In fact, over a twenty-year timeframe—which is to say, during the most crucial window for stabilizing our planet before it's too late—methane traps about *eighty times* more heat than carbon dioxide. Betting on gas as a bridge fuel would send global temperatures soaring. In fact, it doesn't take a lot of leakage to make natural gas even dirtier than coal.

You might ask, "What if we plugged the leaks?" But methane is colorless and odorless, which makes leaks notoriously difficult to spot, let alone plug. Even when you can find them, they usually can't be fixed right away—the Aliso Canyon leak, discovered near Los Angeles in 2015, sent over 100,000 metric tons into the atmosphere before it was stopped up about five months later. And let's say, for argument's sake, that we develop the technology to find and quickly repair natural gas leaks and deploy that technology in the United States and a few other wealthy countries via new infrastructure, in the next four or five years. That's not likely, but it's doable. The problem is, what about the other 80 percent of the world? Are we really going to build a global natural-gas infrastructure that will only be used for five or ten years, all on the assumption that the entire world will be able to contain leaks as well as the United States and Germany can, or that when they happen, the poorest countries will be able to spend the huge amounts of money it takes to fully contain them?

Also, as it turns out, gas isn't even necessary as a bridge to clean energy. If we'd invested heavily in gas ten years ago, when lots of clean-energy technology was still at its earliest stages, that might have made at least some sense. But right now, the quickest way to transition to clean forms of energy is to transition to clean forms of energy. In addition to solar and wind, we can do much more geothermal. We can try to develop smart, safe ways to use nuclear energy. We can build all this *faster* than we can build a global natural gas system. And in most cases, it will cost less, too.

Basically, natural gas is a bridge to nowhere. At best, it's useless, because by the time we're done building it, we're already on the other side. At worst, it speeds up climate change dramatically by leaking enormous amounts of planet-warming methane into the air at a moment when it's never been more important to get to net-zero emissions. If you know how the gas industry works in detail, it's easy to see that going all-in on gas as a bridge fuel is a risky and unnecessary choice.

But most people *don't* know how the gas industry works. And even if they did, and totally rejected the bridge-fuel theory, the fossil-fuel people would just move on to some other talking point that sounds persuasive until you look more closely. That's one reason bullshit can be such an effective strategy for desperate people and desperate industries trying to buy time. Picking apart misleading arguments is exhausting—and there is an endless supply of them. Every day that we are forced to spend debunking bullshit is another day the fossil fuel industry can continue to make obscene profits at humanity's expense.

So what if we tried a different approach? Rather than analyze the fossil-fuel industry's case in detail, and rebut it with technical knowledge most people understandably don't possess, what if we take a step back and try to spot an idea bubble—a closed-off world where everyone shares the incentives, perspectives, and interests?

Suddenly, detecting bullshit gets really easy. Because when you look at all the people making the case for natural-gas-as-bridge-fuel, you notice a strange coincidence. They all own natural gas. In fact, they all stand to make a ton of money if the rest of us buy into their bridge-fuel idea. It's like standing in the park and realizing that all the people trying to convince you to play three-card monte know one another.

Even the most seemingly independent proponents of bridge-fuel have turned out to be not so independent after all. For several years, from the early 2000s until 2012, the Sierra Club endorsed the bridge-fuel argument. At first, that was really persuasive—an environmental group agrees that natural gas is clean! But once people started poking around, they discovered a strange coincidence: Chesapeake Energy, a big natural gas company, was secretly giving the Sierra Club an enormous amount of money.

I'm not saying that the people at the Sierra Club, whom I have a great deal of respect for, were corrupt or trying to mislead—I'm sure they genuinely agreed with the bridge-fuel proponents. And to their credit, once the secret funding was revealed, they changed leadership and backed away from their support of natural gas. But as long as the Sierra Club relied on the natural gas industry for funding, it had incentives to see the merits in the bridge-fuel argument and to ignore its many flaws.

Part of what makes idea bubbles so dangerous is that you can fall into them by accident, and when you do, it clouds your judgment. The taxi medallion bubble in the early 2010s was a perfect example of this. From the outside, the moment Uber arrived, the writing was on the wall. But from inside the taxi industry, everyone had an incentive not to see what was obviously coming. They all benefited when the price of medallions went up, so they found new ways to convince themselves that the price of medallions would always go up.

I've seen this happen in climate, too. Just a few months ago, a friend who runs a climate-focused fund was preparing for a trip to Texas to raise money. "The science looks bad," I said to him, "much worse than we thought it would be."

"You know," he replied, "I don't agree with you."

What I said was, "Really?" But what I thought was this:

Of course you think that! Because you're about to go pitch all these oil millionaires and billionaires on your fund. You can't go there and say, "Hey, if we keep pumping oil and gas at these levels, it'll destroy the planet." You have to say, "No one's perfect, but don't worry, you can keep pumping all the oil you want, and you don't have to feel bad about it. Now can I please have a billion dollars?" I don't think my friend was lying to me, but incentives are powerful things. I think my friend was bullshitting himself. No one bites the hand that feeds them.

For what it's worth, this is one reason why, even though I disagree with people from the fossil fuel industry, I'm always interested in hearing what they have to say. First, their internal climate science has actually been very good—even if they've spent most of their time and energy denying its conclusions—so maybe they know something I don't. Second, if you're going to be in contention with somebody, it's helpful to understand how they're thinking and what they might do next.

That said, there's a big difference between listening and believing. One of the things I learned as an investor is that a surprising number of people take arguments at face value, especially when those arguments are stated confidently. They feel overwhelmed by how much they don't know, or they're too busy to ask questions, or they're just trusting and credulous. Whatever the reason, it makes them easy marks.

During my time on Wall Street, before moving to San Francisco to found Farallon, the easiest mark of all was known as "the Pasta King." I never learned his real name, or even whether he was actually in the pasta business. All I know is that once a year this guy would arrive on Wall Street from Italy with a huge fortune in tow and announce he wanted to buy stuff.

You can imagine what happened. Across Wall Street, investment bankers waited all year for the Pasta King to arrive. They'd make lists of every dud they couldn't unload, the assets that were so obviously terrible no one would buy them. And then they'd sell them to the Pasta King.

You might be thinking, "That guy sounds really stupid!" But the truth is that most of us are the Pasta King at some point in our lives: naïve, unquestioning, accepting the wisdom of the self-proclaimed adults in the room without thinking critically. And when it comes to climate, that's exactly what oil and gas is counting on. They're looking for pasta kings.

Take the opinion section of the *Wall Street Journal*. They're not part of the oil and gas industry, exactly, but as the Murdoch family's flagship American paper, they're a crucial part of the ecosystem that supports the status quo. A 2015 study of several leading US newspapers found that the *Wall Street Journal* was the least likely to discuss the threat of climate change, as well as the most likely to use negative economic framing when discussing climate change solutions.

The *Journal* op-ed writers rarely engage in outright lies. What they frequently do instead is use facts without context in a flagrantly misleading way. For example, for a very long time, their columnists kept pointing out that 1998 was the hottest year on record, and then they would ask some version of, "If climate change is real, why did the world's hottest year occur so long ago?" Here's Holman Jenkins Jr., a member of the editorial board, in 2013: "The fact remains, in all the authoritative studies, the warmest year on record globally is still 1998 and no trend has been apparent globally since then." And here he is again two years later: "When climate reporters robotically insist, as they did again this week, that the 2000s represent the hottest period in the rather skimpy, 134-year historical record, they are merely reiterating that the pre-1998 warming happened. No clear trend up or down has been apparent since then."

Unlike with bridge fuel, you don't need to know much more detail to understand why this argument is bullshit. You just need a little bit of context. Yes, it's true that 1998 was abnormally warm. But nine of the other ten hottest years on record have come in the last decade. To try to be as fair as possible to Jenkins, he was suggesting that some major warming event happened before 1998, and that nothing's changed since. But that was hard to believe even in 2013 and 2015, when those op-eds were written, and since then, unfortunately, global temperatures have increased even further.

The *Journal's* editorial page was hoping that, with their authoritative tone and prestigious masthead, they could cow their readers into not asking basic, commonsense questions: "Why are these guys choosing one highly specific data point, with no other context to support it? And why are they using that data point again and again?"

And it's not just the *Wall Street Journal* that's been looking for a pasta king. These days, when you talk to oil and gas executives, they'll often discuss their commitment to reducing emissions per barrel of oil. They can even break out some impressive numbers to show you how much progress they're making. It sounds pretty good—until you ask a single question. "Is the total number of barrels you drill going to go up or down?" It turns out that their plan for the next several decades involves extracting far more oil per year than they do already, which will more than cancel out any gains they make in efficiency.

It's easy to think, given their reach and power, that the fossil fuel companies are masters of manipulation and spin. But sometimes all it takes to expose their arguments is the willingness to ask the most basic questions. With just a small commitment to maintaining your skepticism, you can avoid being the pasta king—and make your bullshit detector much sharper in the process.

There's another, even easier way to get better at spotting bullshit: instead of focusing on misleading statements, focus on misleading people.

My friend Billy and I once discussed this, in a roundabout way. Billy ran one of the largest independent studios in Hollywood. Before that, he was a big-deal agent, and one day, about twenty years ago, he called me from his desk at the agency.

"Tom, do you use a voice stress analyzer?" he asked.

"Billy," I replied, "I don't know what a voice stress analyzer is. What are you talking about?"

"You know, to see if people are lying."

He explained that a voice stress analyzer is a small device you attach to your phone. When someone calls you, your analyzer picks up the vibrations in their voice, and when they lie, their vibrations go up.

"It's subconscious," Billy told me matter-of-factly. "Everyone in Hollywood uses it."

"No way. You have one of those?"

"Are you kidding me? I wouldn't even think of having a conversation without it."

In case you're considering buying a voice analyzer of your own, you can save your money. They've been proven not to work. But what struck me most about that conversation was Billy's overall approach. He was looking for a foolproof way to separate the verbal wheat from the chaff.

In my experience, there's a much better strategy. Stop trying to figure out where the lies are and start trying to figure out who the liars are instead.

But don't take it from me. Take it from one of the world's great bullshit detectors: Jane Austen.

I'm not kidding. You can spend years trying to invent your own techniques for becoming a better reader of human character, or you can rely on a smart, sensitive writer who peered as deeply into human nature as any of us ever will, and you can learn from her insight and wisdom for less than the price of a Big Mac. Plus, you'll probably feel happy at the end. (Virtue always triumphs with Jane.) That seems like a pretty good deal to me.

My favorite Jane Austen novel is *Pride and Prejudice*. I love the way (spoiler alert) Mr. Wickham seems too good to be true. He's charming, he's got great manners, he tells a moving sob story about how Mr. Darcy treated him badly—and then he turns out to be a total jerk. Meanwhile, Mr. Darcy appears cold and aloof but he's completely trustworthy. By the end, you don't have to evaluate every one of Mr. Darcy's statements to know that you can trust him, just as you don't have to rebut every one of Mr. Wickham's to know he's a liar.

Today, the fossil fuel industry is the climate world's Mr. Wickham, putting on a full-scale charm offensive. The oil and gas companies want you to think that they understand the threat to the planet and that they're ready to change. To decide whether to believe them, you can employ your own version of a voice stress analyzer, going through each greenwashing statement or upbeat ad trying to determine if it's false. Or you can apply the Jane Austen test and ask: "Are these guys trustworthy?"

The record is pretty clear. The fossil fuel companies have lied through their teeth for decades. First, they denied that climate change was happening. Then they denied that climate change was caused by humans. Then they said the science was in question. Then they said warming wasn't happening as fast as independent researchers said it was.

They weren't just wrong. They *knew* they were wrong. They'd read the same research as everyone else. Their own research even confirmed it—in fact, for a long time their climate models were better than NASA's. And they continued to lie anyway. They were willing to put millions, maybe even billions, of lives at risk to protect their profits.

That doesn't mean that everything a fossil fuel person says is guaranteed to be a lie. But it means that oil and gas doesn't deserve to be trusted implicitly. They've been making suckers out of us for way too long, and it's only reasonable to treat what they say as bullshit until proven otherwise. We don't have the time to debunk every talking point made by an industry that's consistently lied through its teeth. Think about the number of years we spent arguing that climate change was real instead of doing something about it. We won that argument. Even the oil and gas companies now admit that the climate crisis is happening. But we lost at least a decade of action, which meant an additional decade of rampant profits for the fossil fuel industry as it pumped more and more pollutants into our atmosphere for the rest of us to deal with.

Climate denial has become climate delay. The dead-enders know that eventually our world is going to change, and that the fossil fuel industry, like American whaling in 1860 or New York taxicab medallion-hoarding in 2010, is going away. We can side with the fossil fuel folks and try to delay the inevitable for as long as possible—with dire consequences. Or instead, we can find, and listen to, the climate Mr. Darcys. I'm lucky to have plenty of them in my life.

One of them is my friend Bill McKibben. Bill is a professor at Middlebury College, an award-winning activist, author of a dozen books on climate, and one of the founders of 350.org, an organization that partners with hundreds of grassroots campaigns around the world to fight the climate crisis. He's even got a species of gnat named after him—*Megophthalmidia mckibbeni*. If I have a scientific question about climate, one of the people I call first is Bill.

Then there's Johan Rockström, an understated Swede who runs the Potsdam Institute for Climate Impact Research. Johan is a scientist, and as far as I'm concerned, he's as good as it gets. He's done primary research. He's pursued truth very hard and very intentionally. He's never tried to puff himself up or blow his own horn. He doesn't pull punches, but he's also not an alarmist. He lays out the evidence. And he's really good at explaining it all in a way that non-scientists can understand.

I don't assume that Bill and Johan will be right all the time. No one is. But I trust them to be knowledgeable and honest. And when it comes to making big decisions about the future, that's what counts.

The clean-energy revolution has already turned a corner. Its momentum is growing stronger every year. That's an amazing thing—but it also means we need to brace ourselves for more fossil-fuel bullshit than ever. Their goal is not really to win an argument. It's to keep the status quo intact for as long as possible. Each year of delay means new levels of warming and extreme weather for our planet. But for the oil and gas companies, it means tens of billions of dollars in new profits. For executives at those companies, it means tens of millions of dollars in salaries and bonuses. For banks, it means avoiding having to write off bad investments. For politicians, it means avoiding a reckoning with the voters whose future they're willing to throw away.

Climate people don't have to parry every attack or misleading claim. Instead, we need to ask some basic questions: *Is this person coming from a bubble of people who think the same way, and who stand to benefit by not thinking differently? Am I the pasta king, or have I peered below the surface of their claim? Can I trust them, or have they lied before?*

Once we do that, we can stop getting bogged down by bullshit and start paying attention to the people who really deserve our trust, before it's too late.

CLIMATE PEOPLE

Harold Mitchell

A s we've seen in recent years, no one is immune from the effects of climate change. But lower-income communities are disproportionally harmed.

A hotter planet creates fewer work opportunities in outdoor industries like farming and construction, which employ roughly 10 million people in the United States alone. People who live in homes with older infrastructure, and are less likely to have reliable transportation, are at the greatest risk from severe weather events caused by a warming planet such as hurricanes and wildfires. A hotter planet also means more air pollution from greenhouse gas mixing with dust and aerosols, leading to more asthma attacks and respiratory infections, diseases found at higher rates in lower-income communities.

Harold Mitchell knows about environmental injustice firsthand. And he's dedicated his life to fighting it.

Harold grew up in the Arkwright neighborhood of Spartanburg, South Carolina. When he was a kid, about 96 percent of the neighborhood's residents were Black and a quarter lived below the poverty line. For decades, the city operated an open landfill in Arkwright, while just a few hundred feet away the International Minerals and Chemical Corporation used the neighborhood as a dumping ground for waste from a fertilizer plant.

Harold began experiencing health issues in his twenties, just as his fifty-nine-year-old father was diagnosed with terminal lymphoma. After learning that people throughout Arkwright were falling ill, Harold worked with local groups to demand answers. The EPA eventually agreed to look into the fertilizer plant and dump site and found seventy pollutants, thirty of which were in concentrations three times the level considered harmful to human health. They declared both areas Superfund sites and ordered the waste removed.

For some people, that would have been enough. But Harold recognized that the problems caused by pollution were just one symptom of a wider environmental neglect of his city, and in 1998 he founded the ReGenesis Project, a nonprofit working to revitalize Spartanburg. The oil and gas industry and its allies often suggest that environmental justice means stopping economic development from going forward. But Harold and ReGenesis disprove that every day. For them, environmental justice means moving forward in a better way, making sure that those without the loudest voices aren't taken advantage of by wealthy and powerful corporations.

In the two decades since its founding, ReGenesis has helped clean up groundwater pollution, build a green recreation center, and construct new parks and green spaces while bringing new health centers, grocery stores, and hundreds of affordable homes powered by solar energy to Spartanburg. Today, ReGenesis is working to turn the old fertilizer plant into a sustainable hydroponics and aquaponics facility, where nutrient-rich water that comes from farming fish is used to grow plants, which in turn filter the water for the fish.

They're also working on creating a factory that will produce 100 percent recycled and recyclable fabric. This will create jobs for the community—and since mass production of clothing that ends up in landfills sends large amounts of greenhouse gas into the atmosphere, it will help us get to net zero, too. And they're not stopping there. Perhaps most important of all, Harold recently expanded the ReGenesis Project into the ReGenesis Institute, which aims to help underserved communities across the country replicate Spartanburg's success.

Harold and ReGenesis are showing that when it comes to climate, justice and prosperity can go hand in hand. In fact, over the long term, the second depends on the first.

CHAPTER III

KNOW WHAT TO KNOW

The fossil fuel industry loves to point out that it's impossible to know everything about climate change. Until fairly recently, their angle of attack, developed in part by Republican messaging guru Frank Luntz, was to question the science regarding the very existence of global warming. Sure, 97 percent of scientists said climate change was real and caused by human activity, but what about the 3 percent who didn't?

This talking point was always misleading. For decades, it's been impossible to find a serious, respected climate scientist who didn't agree that the climate was changing, and that humans were, by far, the most likely cause of that change. Today, even Frank Luntz has, to his credit, admitted he was wrong on climate. The oil and gas companies haven't made a similar admission and apologized for their decades of climate change denial, but as climate change has gone from prediction to reality, they've stopped casting doubt on the consensus that greenhouse gas emissions are warming the planet and changing the weather.

Still, even though the fossil fuel industry is no longer doubting the existence of climate change, they continue to make a version of the same basic argument they did starting in the 1990s. They'll point out that we can't be exactly sure which clean-energy technologies will work best, or what the weather will be like on any given day in the future, or precisely how geopolitics will affect other countries' efforts to cut greenhouse gas emissions. All that is true—when it comes to making predictions about the future, it's impossible to know everything about anything. But the conclusion that they draw makes no sense: "We don't know everything about climate change with 100 percent certainty, so we shouldn't take any meaningful action."

The fossil fuel industry is pointing to, and trying to benefit from, something very real: for human beings, not knowing is unsettling. Staring into the face of the fundamentally unknowable is, for most people, one of the most emotionally uncomfortable experiences we can have. Many people become paralyzed while they wait for perfectly complete information, which never comes. But imagine if you made other decisions in your life the way the fossil fuel industry wants you to make decisions about climate. If you only saw a movie if there was a 100 percent chance you'd like it or took a new job only if there was a total guarantee you'd find it fulfilling, you'd miss every opportunity.

Not knowing is particularly stressful during a crisis. During my investing career, I've watched so many people completely lose their cool when the market started to fall apart and it became clear that things were going to get really bad. Everyone has the urge to panic, myself included. But you have to figure out how to resist it. Because once you start panicking, you stop taking in and considering new information. I've seen it happen over and over again. People make decisions that they may even tell themselves are rational. But what they're really thinking is: "I can't handle not knowing what's going to happen. I need to solve this problem. I need to end this pain. I need to do something, anything, immediately." When you're that desperate to regain a sense of stability, you stop thinking about whatever problem you were trying to solve. In investing, it's what drives people into a fire-sale mentality. People will sell assets they know they shouldn't sell, just so they can stop worrying about how things will turn out.

But with climate, as with other difficult, high-stakes decisions, we have to make our choices without being able to know exactly how the future will turn out. Waiting until we know everything there is to know would mean giving up, condemning ourselves to a world defined by climate catastrophe.

What we need, then, is a better way to handle the fundamentally unknowable. In my experience, such a way exists—it just requires a change in focus. Instead of asking ourselves, "How much do we know?" we need to ask a different question: "Do we know *what* to know?"

I first learned the importance of asking this question on a ranch in Nevada. The summer after my freshman year at Yale, most of my classmates headed for résumé-padding jobs in New York or study-abroad programs overseas or tennis-coaching jobs at their country clubs. I went to Gardnerville, a town of approximately 3,300 about an hour south of Reno. Sallie Springmeyer—a very smart, sophisticated woman who had gone to law school and been married to the state's district attorney—owned a ranch there. She'd put up a flyer at Yale looking for ranch hands, and I'd answered it.

I wasn't brand new to the outdoors. My grandfather on my mother's side was a hunter and fisherman who briefly lived on the Blackfeet Indian Reservation in Montana. My mom rode horses, and she fished and hunted just like her father did. In the summers when I was a kid, we would spend two weeks each year visiting my grandparents on a lake in Minnesota, and my cousins and brothers and I would fish every day—bass, perch, crappies, walleye.

But that wasn't really why I wanted to spend a summer as a cowboy. It might sound corny, but at least it was sincere: I wanted to see America. I wanted to live and work someplace as far from Manhattan as it was possible to be, at least figuratively speaking.

I got my wish. I spent that summer in a hut with no running water, a galvanized tin roof, and wood walls so full of holes they were essentially see-through. The only bathroom was an outhouse. Every morning before breakfast, I'd get up and milk two cows. Then I'd spend my day moving cattle, fixing fences, bucking hay, or whatever else needed to be done.

A bunch of real-life cowboys worked on that ranch. Some of them I liked a lot. Others I was kind of freaked out by. My co-workers on the ranch knew where I came from, and where I was going back to in the fall, so they enjoyed putting me in my place. One day, my job was to catch a bunch of male piglets so that a ranch hand could castrate them. I was thinking there would be an operating table involved, maybe some anesthetic. Not exactly. My "operating room" was a square pen, about ten feet per side, occupied by a 500-pound sow and twelve baby pigs. About half of them were males, and I had to hunt them down, grab them, and hand them to the guy so he could do the surgery with his folding knife. No anesthetic, and only an ointment called "Bag Balm"—a salve designed to soothe udders on cows—to prevent infection.

That day, I learned that when a pig gets castrated, it screams exactly like a two-year-old child. And what does the mother do? She tries to kill whoever is threatening her children. So there I was, in a pen with an enraged 500-pound sow, trying to dodge her while scooping up her remaining piglets. Meanwhile, the cowboy stood outside, thinking it was the funniest thing he'd ever seen in his whole life—this kid from Yale trying to evade an infuriated sow. In fairness—and with several decades of hindsight—I can see why.

Somehow, I made it out of that pen alive. What's more, I enjoyed the experience on the ranch so much that the next summer I went to Oregon to pick cherries—a different version of looking for America. My co-workers were mostly migrant laborers from Mexico rather than cowboys. But in many respects, not least the sheer physicality of the work, the experience was similar.

I learned three big things over those summers. The first is that there's no such thing as unskilled labor. Our free-market society values some skills more than others, but any time I hear someone describing back-breaking work as "unskilled," I think back to the pickers on the orchard who could collect three bushels in the time it took me to collect one, or to the ranchers who could expertly corral a dangerous young bull. *You think that's unskilled?* I want to say. *Why don't you try it?*

The second lesson I learned on the ranch and in the orchard is that there are some things in life you just can't know.

I'm not talking about metaphysical questions like *What is the meaning of life?* or *What happens after we die?* I'm talking about the here and now. What I saw during those two summers was that the web of relationships within nature—among animals, people, weather, sun, soil, streams, fruit, livestock—is far too vast and complicated to fully comprehend. And furthermore, it's impossible to bend it to our will.

Some people I worked with couldn't stand the idea that nature would always be beyond their control. The ranch foreman, whom I'll call Doug, "trained" horses by hitting them in the face with a shovel. He refused to believe he couldn't bend a wild creature to his will. Not surprisingly, his tactic backfired. Almost every horse he trained was terrified of people. If you shifted your weight ever so slightly in the saddle, or shuffled your feet in the stirrups, or did anything else even the tiniest bit surprising, the horse would freak out and buck.

It was clear to me even back then that Doug's reaction to his fear of the unknown wasn't productive. Rather than balk at the idea that some things were unknowable, you had to learn to accept it. If you wait for perfect information, you'll be left waiting forever. If you act as though you understand something completely, you'll make bad decisions.

At the same time, sometimes if you zoom out far enough and try to understand the big picture, you can find something knowable and predictable—some essential truth you can act on. In my case, after my summer on Sallie Springmeyer's ranch in Gardnerville, I went home and bought ten shares of stock each in John Deere and International Harvester.

When I think about it now, this was in some ways a foolish investment. I knew what a Deere tractor was. I knew what an International Harvester combine was. But I knew nothing about the companies from a business perspective. What were their market caps? How effective were their CEOs? Did they have a lot of bad debt?

But in another important respect, my early stock purchases were sensible. After my summer on the ranch, it became clear to me that agriculture was a business that depended on a million inscrutable variables. But I'd also noticed that no matter how those variables turned out, in good times and in bad, ranchers and farmers were buying new equipment and repairing old equipment.

In other words, while agriculture was inherently prone to wild and seemingly random fluctuations, selling agricultural *machinery* was much more consistently profitable. The way the companies that made these machines sucked up all the available cash in the system wasn't remotely fair to farmers, but the fact remained. If you knew what to know, you could make money investing in agriculture.

The same basic idea applies to climate. We may not know everything, but we can know a few big things. What's the climate equivalent of investing in agricultural machinery rather than buying livestock or growing crops? The answer is pretty simple. I'd say we should look at the basic relationship between greenhouse gases like carbon dioxide and methane and the gradual warming of the planet. If you know the basic, predictable fact that more emissions equal more disruption to our weather patterns and, by extension, to humanity, then it's not that difficult to figure out, broadly speaking, where to invest your time. It's also why getting to net zero is such an important goal. Regardless of how we reach that point, we know that if we get there, we'll have stopped making things worse and can start to make them better.

Another version of knowing what to know is what I think of as "drafting a quarterback." If I tell you that a football team has a once-in-a-generation talent at wide receiver, there's still a lot you need to know before being able to guess how the team as a whole will perform. But if I tell you that a team has a once-in-a-generation talent at quarterback, it's a different story. You may not know everything, but that one piece of knowledge is enough to tell you that the team is a contender.

In the investing world, a lot of people struggle not because they're bad analysts but because they're analyzing the wrong things. They're focused on the wide receivers when they should be trying to draft a quarterback. Here's a real-world example. If you wanted to make a lot of money in the stock market over the last thirty years, you didn't need to form tons of small ideas about which companies would do well and which wouldn't. You needed just one big idea: everyone is buying more software and information technology, so it's smart to invest in companies that have really strong IT. That idea—and that idea alone—would have led you to everything from Amazon to Walmart to Uber to Airbnb. Knowing that one piece of information would have been like having Tom Brady on your team.

One of my favorite examples of the difference between quarterbacks and wide receivers comes from my experience in Hollywood. You may not think *The Usual Suspects* (neo-noir mystery), *American Pie* (raunchy teen comedy), and *Topsy Turvy* (critically acclaimed small-budget period musical) have much in common. But in fact they were all financed by my firm. From the mid-1990s to the mid-2000s, Farallon Capital was one of the largest independent funders of films in the world.

Despite being involved with blockbusters across a wide range of genres, I freely admit that I don't know anything about how to make a good movie. Lots of people get involved in the entertainment business because they believe that, deep down, they're Steven Spielberg. I'm not Steven Spielberg. I have a lot of respect for filmmakers, and one measure of that respect is that I know I'm not one of them.

So, what gave me the confidence to invest in films? Two British guys came to us with a quarterback of a business model.

It worked like this. At the time, there were just six big film markets in the world: the US, the UK, France, Germany, Japan, and Korea. (It's remarkable, considering the state of the film industry today, that China wasn't even on the list.) If you could package together a script, a director, and the stars, you could sell the rights to distribute that movie in some or all those markets. You could then use these foreign "pre-sales" to cover the costs of making the movie. Once the box office opened, the upside would come from the markets you hadn't yet sold the rights to, from sharing a percentage of the distributors' profits, or both.

Think about that: we hadn't shot a single frame. We hadn't scouted a location. But we still had a guarantee that if the movie got made, we would at least break even. Even better, we could go to an insurance company and pay a small fee to "bond" the movie—meaning that if the movie *didn't* get made, we'd get our money back. The upshot of all this was that for no money down, you could guarantee yourself a 6 to 8 percent return, and with any luck make it 20 percent.

I love good movies—but in our business, we thought of movies as basically another form of software. Part of the secret to our success was that we left the creativity to the creative people. We wanted to back hit movies, of course. But in Hollywood—the business where, famously, nobody knows anything—it's impossible to guarantee a hit. We didn't try to study the million factors that make a movie succeed or fail. Instead, we drafted a quarterback: we figured out one crucial element of the way movies were financed that turned everything else into icing on the cake.

Like a lot of businesses, our film financing model worked well until it didn't. Competition increased. The movie business changed, and foreign rights for independently financed films became harder to sell. But another issue was entirely of our own making. Our two British partners were great—until they started to think they were Steven Spielbergs and decided they could figure out the secret to producing huge hits. They forgot about the quarterback and focused entirely on wide receivers. Not long after, our performance started to decline.

When it comes to climate, I'd argue that drafting a quarterback means focusing on decarbonization. We can't know for sure which types of new technology will have the greatest long-term impact, or provide the most energy at the lowest cost at some point in the future. But we don't have to. As long as we're doing whatever it takes to bring down the overall amount of carbon (and methane) that we release into the atmosphere, we'll be moving in the right direction. And if we can bring those pollution numbers down to zero, and fast, we'll have turned the corner on the biggest threat facing humanity today.

That's also a helpful way to think about your own part on climate. There are all kinds of actions that you, personally, could take right now. You'll never know for certain which of them will be most effective or most worth your time and attention. But just as so many members of my parents' generation asked themselves, "What can I do to give the Allies the best chance of winning the war?" a good question for climate people is, "What can I do that has the biggest impact on the world's greenhouse gas emissions?" Throughout this book, I'll highlight people and strategies that I hope help you find your answer. But the mere act of asking the right question will help you invest your time in the best possible way.

But what kind of action can we, as individuals, take? How can we be part of the global effort to stabilize our planet as quickly as humanly possible?

Fortunately, figuring out how to reduce carbon pollution isn't one of those unknowable mysteries of life. If humanity is going to get to net zero, and if you, as an individual, are going to do your part, a good place to start is to consider what I think of as a "five plus one" approach—five areas where we'll need to cut our emissions, and one where we can undo the damage from the emissions we do create.

Even if all you know about solving the climate crisis is the information contained in these six paragraphs, I suspect you'll be able to begin finding your role in how you can help humanity get there.

1. **Electricity generation.** The great thing about fossil fuels, if you ignore the fact that they're killing us, is that they're powerhouses of stored energy. For more than a century, we've turned that pent-up energy into heat and used that heat to create electricity. We need to create more electricity than ever as things that used to be powered by fossil fuels, like cars, are going to become electric; as the world population grows; and as more countries access the miracles of modern life. If we want to do this without destroying human society as we know it, we need to create all that electricity in a way that doesn't rely on fossil fuel–generated heat. That means building up existing clean and renewable technologies like solar, wind, geothermal, and, if it can be done cleanly and safely, nuclear. It means trying to find new sources of energy and electricity, for example through nuclear fusion. It also means that we need to support all the things that get the electricity wherever it needs to go—better batteries, transmission lines, and power grids.

2. **Transportation.** Burning fossil fuels is one way to move vehicles from Point A to Point B. We need a better way. Electric cars, electric trucks, electric buses, and the charging networks needed to connect them to clean-energy grids will play a large role here. Making public transportation more efficient and reliable will help

get cars—and the pollution associated with manufacturing them as well as driving them—off the road as well.

But other types of transportation are harder to electrify—especially passenger and cargo planes, because at the moment, any battery big enough to power them would be too heavy to get off the ground. Shipping—whether by truck or boat—is hard to electrify as well, for a similar reason: a big battery takes up space that could be occupied by cargo. That means we'll need to rely not just on new sources of electricity but on new battery technologies and new sources of fuel. (In my mind, the most promising of these is green hydrogen. We need to move down the technology cost curve, but if we can make it work, it will be better, faster, cheaper, and cleaner than anything else out there.)

3. **Manufacturing.** Let's say you buy an electric car and hook it up to a grid powered by renewable energy. That's zero greenhouse-gas pollution. But what about all the carbon and methane emitted when they were *building* that car? Or your phone? Or your clothing? Or basically anything else manufactured, which these days is almost everything? The basic problem is that manufacturing takes an enormous amount of energy, and in particular requires very high heat. We need to figure out how to make things using less heat, and how to generate heat without generating as many emissions.

4. **Agriculture.** As we feed a growing worldwide population, the ways we raise plants and animals have to change. For example, nitrogen fertilizer—which didn't exist 120 years ago but today is considered essential to industrial farming—is a huge source of emissions because it's made using fossil fuels. There's also the problem of deforestation. When an acre of rainforest is cut down to clear land for a palm oil plantation or a cattle ranch, it's like wringing a giant sponge full of greenhouse gas into the air. We have to start producing the food we need without all the emissions.

5. **Buildings.** Plumbing. Air conditioning. Heating. Whether it's a house, an office, or any other type of modern structure, our buildings use an enormous amount of energy, and right now that means they create an enormous amount of greenhouse gas pollution. Even worse, most buildings leak. We need to make sure that what we're building today is net-zero emissions. But because it's estimated that in developed economies, about 80 percent of buildings in use today will still be in use in 2050, focusing on new construction alone isn't enough. We need to retrofit old buildings so that they waste less energy—and cost their owners less money in the process.

Which brings me, finally, to the plus one: **sequestration**. The first five areas on this list are about releasing less greenhouse gas into the air. Sequestration is about taking greenhouse gas out of the air. That can involve mechanical solutions like direct air capture—basically vacuuming carbon and storing it someplace it can't escape. It can also involve natural solutions, such as planting trees or kelp beds that absorb carbon, or switching to farming practices that store more carbon in the soil.

Offsets are a valuable tool for getting closer to net zero. I buy them myself, particularly when I travel. Still, I'm always very careful when I talk about sequestration, because it's not a magic bullet, and it almost certainly never will be. Sequestration might well someday be the most effective option for a few types of emissions that, combined, make up an estimated one-quarter of our current total. But in most cases, cutting pollution is a lot more realistic, and a lot less expensive, than trying via sequestration to cancel it out.

One big problem with sequestration is that unless we're careful, it can become a buzzword, a way to pay lip service to action while doing little or nothing at all. For example, if I pay to plant an acre of trees, think about all the questions I need answers to before I know how much carbon I'm offsetting. Is a real live person planting seedlings in the ground or is a guy in a helicopter dropping a bunch of seeds out the window? Were the trees going to be planted anyway? How do I know the company selling me my trees isn't also selling them to someone else? How big will the trees grow, and how quickly—in other words, how much carbon will they be able to sequester? What are the odds that in the next ten years, on our increasingly dry and hot planet, my trees will burn down, releasing their stored carbon back into the atmosphere?

It's possible to answer these questions and confirm that offsets are genuine, permanent, and additional. But the incentives to game the system are high—especially because offsets that work tend to be much more expensive than so-called offsets that make you feel good but don't help us get to net zero.

If we're not careful, offsets—and sequestration more generally—could trap climate people in our own idea bubble, one in which we think we're helping to stabilize the planet when in fact we're just maintaining the status quo, which is exactly what the fossil fuel industry wants. A series of studies have found that most carbon-offset projects haven't reduced emissions by anything approaching the levels they claimed. Similarly, while I'm all for reforestation, the brutal fact is that it will take decades of planting trees to offset a meaningful fraction of our current carbon emissions, and we don't have that kind of time.

Other types of sequestration unquestionably work—but they're very, very expensive. The most talked-about of these is direct air capture, or DAC. Today, the price of sucking a ton of carbon from the air and storing it somewhere safe is somewhere around $600 per ton. Another way to say this is that it would cost about six dollars to remove the carbon emissions generated by driving on one gallon of gas. On the internet, you'll find people saying that we can bring the price of direct air capture down to $15 to $25 a ton. No way. Professionals involved in the technology think *maybe* we can get the cost down to about $130 a ton (or a little more than a dollar per gallon of gas), but even that will be very difficult and take decades to accomplish.

A few other types of sequestration are quite promising. For example, I've invested in a company that does what's called "enhanced rock weathering," basically drilling tiny holes in certain types of rock so that they naturally absorb more carbon. But even if all the sequestration methods, new and old, pan out—and could cancel out pollution in a cost-efficient way—it's expected that they'll be able to pull 10 billion tons of carbon per year out of the atmosphere. Last year, global carbon emissions were 42 billion tons.

Sequestration can certainly be part of the solution, by canceling out the greenhouse gas we can't avoid sending into the air. But anyone who says that sequestration can get us to net zero—that we can keep emitting all the planet-warming pollution we want and simply cancel it out—is either wildly ignorant, naïve, or lying. So it shouldn't surprise you to learn that the fossil fuel industry's plan for fighting climate change includes treating sequestration as the answer to all our problems—so much so that oil and gas lobbyists basically forced Congress and the Biden Administration to spend taxpayer dollars researching carbon capture technology.

It would be easy to say that oil and gas has no plan whatsoever for what to do as the planet warms, but that's not quite true. In presentations to investors, and increasingly in public, they've made it clear that they have a climate-change strategy of their own.

Step one is to extract as much oil and gas as possible. More drilling. More pipelines. More pollution. More profits for the executives at the fossil fuel companies and the bankers who lend money for new oil and gas projects. Of course, if we extract and burn all that oil and gas, we won't just fail to lower our carbon emissions—we'll increase them by huge amounts. For the most part, the fossil fuel industry's attitude is, "Let's not have a plan. Let's just hope we do a little better every year."

That just won't work. They know that what they're doing will likely result in a global human tragedy, but they believe their job is to make money, not save the planet.

Step two is to try to sequester all the carbon we emit, even though for all the reasons I've discussed above, that's clearly impossible. Instead of a five-plus-one approach, the fossil fuel companies want to focus entirely on taking a virtually unlimited amount of carbon pollution out of the atmosphere, because that would allow them to burn a virtually unlimited amount of fossil fuels. If someone can come up with a magic-bullet technology that allows us to get carbon out of the atmosphere cheaply in enormous quantities, that would indeed be terrific. But it's very unlikely to happen.

Here's just one example of the oil and gas strategy at work: fossil fuel companies are putting a lot of energy, or at least a lot of marketing, into the idea that captured carbon and methane could be pumped back into old oil and gas wells to be sequestered there. It sounds like a perfect, green solution—or at least it might seem that way to the Pasta King. The rest of us would probably ask, "What happens if there's a leak in one of those wells?" The answer is that an enormous amount of greenhouse gas would be released all at once. And there's another bigger problem with the idea of filling old wells with captured carbon. CO_2, which is a gas, takes up much more space than oil or liquid natural gas, which are (obviously) liquids. As my friend Hal Harvey says, putting carbon emissions back into wells would be like stuffing a piece of popcorn into a space the size of a popcorn kernel. And turning gas into liquid—which is the most common way oil and gas companies talk about solving this problem—is too expensive to work at scale.

Like so many "green" ideas being pushed by the fossil fuel companies, it's very unlikely to work. And they probably know it. The real reason they're promoting it isn't because energy industry officials don't understand that gas occupies more volume than liquid—it's because it allows them to appear as though they're doing something, when in fact they're just buying more time. Yet again, climate delay has become the new climate denial.

Which brings us to step four of oil and gas's strategy: geoengineering. As I've said before, when it looks like things are falling apart, people start to panic. The fossil fuel industry's plan is to wait until panic sets in and then advocate for desperate measures, like triggering artificial volcanic eruptions to cloud up the atmosphere, spraying salt particles onto clouds to try to turn them into a shield from the sun's rays, or launching giant mirrors into space. These aren't even real technologies yet—they're just theoretical. Researchers are only just beginning to work hard on them. Testing them in the real world, on a global scale, is far from guaranteed to work and quite likely to have massive, unpredictable side effects.

Relying on geoengineering to undo climate change isn't just needlessly risky; it's hugely arrogant. It's the planetary equivalent of trying to train a horse by hitting it in the face with a shovel. If there's one thing we've learned in recent decades, it's that nature reacts to human behavior in ways that are impossible to fully anticipate, let alone calculate. Think about the hubris involved. *We didn't really understand how climate works, so we pumped all this carbon into the atmosphere. But now we'll try a whole bunch of wild experiments on the natural world without any side effects.* The consequences of planetwide geoengineering may turn out to be better than unchecked climate catastrophe, but most likely, it will just be trading one crisis for another.

The climate is a vast, complicated, endlessly surprising system. But the basics of climate change are not. Armed with only the most essential information, and nothing else, anyone can understand the choice each of us faces, see through the fossil fuel companies' strategy of lies and delay, and start figuring out how we, as individuals and as a community, can do more to stabilize our planet—not just in the distant future, but today.

CLIMATE PEOPLE

Benjamin Slotnick

F ew things are more important to a healthy planet than healthy oceans. Through photosynthesis from the plankton living below the sea, oceans create nearly half of the earth's oxygen. They're home to roughly 90 percent of the world's wildlife and support millions of jobs in the fishing and tourism industries. And as a kind of natural sponge capable of absorbing massive amounts of carbon, they serve as some of the most important resources in the fight against climate change.

"Our oceans already remove up to 30 percent of global human emissions," explains Benjamin Slotnick, a geologist and ocean researcher who founded Texas-based Lillianah Technologies. "Not even human ingenuity in most high-technology sectors can approach their solution scale."

Benjamin previously worked in carbon capture, including on a project that stored carbon under the North Sea. But concerns over whether carbon capture was scalable enough to adequately protect the planet compelled him to switch gears, and his background in oceans led him to a new calling: using phytoplankton, also known as microalgae, to bring the ocean's so-called "dead zones" back to life.

Dead zones are created when excess nutrients—often nitrogen or phosphorus from agricultural or factory runoff—flow into water and cause rapid, extensive algae growth, known as algal blooms. When the algae die and decompose, they lower oxygen levels and cause wildlife to die out. Dead zones can turn flourishing marine ecosystems into barren stretches of water seemingly overnight. More than 400 of these dead zones exist today, collectively totaling an area at least as large as the United Kingdom.

Lillianah puts silicate-based phytoplankton into water near dead zones, where they multiply and eventually outcompete the existing algae. Phytoplankton naturally absorb large amounts of carbon, and because they sink when they die, this carbon then becomes stored in sediment at the bottom of the sea. It's an efficient, significant way to get carbon pollution out of the atmosphere. Because they get rid of the original algae blocking sunlight, Lillianah's phytoplankton are also able to help dead zones recover. This helps restore the local populations of animals like fish, crabs, and oysters, ultimately supporting the other animals that make up the massive ocean food chain.

Lillianah's first deployment of phytoplankton was in a dead zone in the Gulf of Mexico, off the coast of Louisiana. It's been successful, removing roughly 100 tons of carbon from the atmosphere while aiding local wildlife. What's more, Lillianah's scientists estimate that deploying more phytoplankton in Louisiana waters alone could remove 4 million tons of carbon dioxide from our atmosphere every year, the emissions equivalent of nearly 900,000 passenger vehicles.

And that's just one dead zone. With oceans covering about 70 percent of the planet, Lillianah has the potential to play a major role in our collective effort to solve climate—and it's this impact that continues to motivate Benjamin and his team.

"The work that we do today," says Benjamin, "will preserve our planet for future generations."

CHAPTER IV

STOP ROOTING
FOR THE END OF THE WORLD

I n September 2023, a back-and-forth between Secretary of Transportation Pete Buttigieg and California Congressman Doug LaMalfa turned an otherwise routine House Transportation and Infrastructure Committee hearing into national news.

While debating the merits of electric vehicles, Secretary Buttigieg said, "Climate change is real, and we've got to do something about it." This should have been uncontroversial. For Congressman LaMalfa, however, this was a chance for a gotcha moment. "Yeah, this one's called autumn," he said. "This climate change right now is called autumn."

Secretary Buttigieg pointed out what just about anyone with a middle-school command of the English language would know: a changing climate is, in fact, different from changing seasons. But in some ways, treating Doug LaMalfa as though he's ignorant was giving him too much credit. In his northeast California district—a major producer of almonds, fruits, olives, and rice—farming is a large part of the local economy. In fact, LaMalfa's own campaign literature describes him as "a fourth-generation rice farmer and businessman."

If that's true, then he's seen firsthand how climate change has made farming more difficult. Hotter temperatures have forced California farmers to change crops they'd planted for generations, and floods have wiped out hundreds of millions of dollars' worth of crops in recent years, sometimes ravishing entire harvests. Not every farmer agrees on what we should do about climate change—but it's hard to find a farmer (let alone a fourth-generation one) who hasn't noticed that the climate is changing.

Flooding isn't the only natural disaster that's caused issues in LaMalfa's district, either. The 2018 northern California Camp Fire, which at the time was the deadliest and costliest wildfire in the state's history, started in Butte County—which he represents. The blaze killed 86 people, caused $16 billion in damage, and burned more than 153,000 acres of land. One of the key factors that made the fire so devastating was a significant drought, with some affected areas having gone more than six months without a single rainstorm. It's hard to imagine that Congressman LaMalfa is the only Californian who hasn't noticed that these kinds of droughts—and the fires associated with them—have become far more common in recent years.

So why is a United States congressman willing to sound so proudly and extraordinarily foolish on national television? Why, for that matter, were so many right-wing media personalities eager to portray him as a hero for doing so?

The answer, I think, is well-summarized by a quote from Upton Sinclair: "It is difficult to get a man to understand something when his salary depends on his not understanding it." That's true not just for money, but for power and influence as well. Doug LaMalfa's standing as a Republican in Congress depends on him denying climate change, so he denies it. A variety of right-wing media outlets and pundits have built their reputations (and in some cases, their funding) on parroting oil and gas talking points, so they keep on cheerleading for their team.

This kind of tradeoff—tying your personal success to others' suffering—is all too common. About twenty years ago, just as people were really waking up to the harm the cigarette companies were causing and beginning to hold them accountable for decades of lies, a friend mentioned to me that he owned a lot of stock in Philip Morris, one of the world's largest tobacco companies.

"Doesn't that worry you?" I asked. No company is perfect, but it seemed to me that cigarette companies belonged in a different category. Yes, the profit margins were high and the cash flow was steady, but the products addict and kill people. I couldn't see any reason, personally, to invest in cigarettes.

My friend disagreed. Not only that, he acted like I was the investment equivalent of a prude. "It's a good business," he said. "They're not breaking any laws. It's not my job to get involved in questions of right and wrong."

Today, a version of the conversation my friend and I had twenty years ago is taking place around the country and the world—only instead of being about cigarettes, the focus is often climate. Politicians attack investors who divest from fossil fuels as "woke" and promise to go after "Environmental, Social, and Governance," or ESG. Like my friend, many of these politicians suggest that it's somehow irresponsible to take anything other than making money into account when it comes to business. In Texas, a group of state legislators even tried to make it illegal to consider climate change when making investment decisions.

First off, the implication that making money is incompatible with doing things that help the planet is totally incorrect. It's true that twenty years ago, there weren't many people making lots of money by investing in clean energy technology and decarbonization, but that's changed. (If anything, Republican politicians ought to be especially quick to recognize this. Elon Musk, a darling of the right, is an electric-car entrepreneur whose flagship company is now based in the Republican-led state of

Texas.) Just as lots of people got rich helping the world transition from whale oil to petroleum, a lot of people are becoming wealthy helping the world transition from fossil fuels to the clean sources of energy that will come next.

What's more, in the coming years, companies that take climate change into account are going to do much better than those that don't—not because they're "woke," but because businesses that ignore major global trends do so at their peril. Barring companies from considering climate change, as some Texas lawmakers recently tried to do, is like requiring them not to think about the stability of global supply chains or the demographics of their customers. It's bad for business.

But the bigger problem with our current political debates over ESG is that they miss the point. The question about whether corporations have an obligation to society, and how they should balance that obligation with their profit-driven mission, is complicated. But it's not the right question.

The real question is: how do you want to spend your life? When you look at it that way, the real dividing lines become clear. On one side, there's a group of people who don't care what the cost to society of their actions is, as long as they make money and don't go to jail. What's especially strange to me is how smug some of these people are—investors who congratulate themselves on their brilliance for buying shares in gun manufacturers after mass shootings, or acquiring companies just to fire half the people who work there and drive up their profit margins.

Adding to human misery may sometimes be good business in the short run, but it's a dumb way to spend your life.

I'm not against making money—I've made a lot of it in my life. But there's an expression my parents used: "money mad." It captures not just greed but the insanity of greed. There's something tragic about letting the pursuit of more wealth disconnect you from the things in life that matter. If you divorce yourself from reality, pretending you're an Ayn Randian superhero who gets to screw over the little people because you're better than they are, I find that offensive. What's more, if you're too boring and uncreative to figure out how to make money without screwing over the most vulnerable and destroying the earth, then maybe you're not the business genius you think you are.

Though I'm not Catholic, I deeply respect two great precepts of the Jesuits: Take care of the world's most vulnerable people, and preserve God's earth. I'm not saying you have to be Mother Teresa. But if you do the *opposite* of those two things, you're probably doing it wrong.

I've often heard money-mad people try to justify their behavior by saying, in essence, that they have no choice—they're just doing what the free market demands. That's ridiculous. If you steal a loaf of bread to feed your starving family, okay, maybe you're Jean Valjean. But if you get out of your Mercedes convertible in your Gucci loafers with your fancy-ass slacks and your Armani shirt and fire everyone in the bakery, you're not Jean Valjean. You're just a jerk.

The way I see it, there are enough things to do that both make money and that you can be proud of. I'm a lifelong investor and an unapologetic capitalist, but there have always been lines I wasn't willing to cross—not because crossing them was illegal, but because being the kind of person who crosses those lines is not a creative or interesting way to live.

Which brings me back to the crucial question so many money-mad people fail to ask: How do you want to spend your time on earth?

And where climate is concerned, there's another version of that question, one that's even more urgent—and that most of us, if we're being honest, might find uncomfortable.

Are you rooting for the end of the world?

In considering the nature of complicity—and the importance of doing something meaningful with one's life—it's hard for me not to think back again to my father's days as a young lawyer in the Navy, helping to prosecute Nazi leaders during the Nuremberg trials. That experience, and the horrors of the Holocaust and of World War II, seared something into him and so many others of his generation: the knowledge that atrocities are not just the work of comic-book villains or deranged megalomaniacs. They're committed by people who find ways to justify their actions, to tell themselves that what they're doing is inevitable, or good, or necessary.

Take Oskar Gröning, the so-called "bookkeeper of Auschwitz." After escaping justice for decades, in 2015 he was finally tried in Germany for being an accessory to hundreds of thousands of murders. The former SS sergeant admitted he knew what was taking place at the camp but said that decades of propaganda had left him convinced that mass killings were, essentially, the cost of doing business in a time of war.

Gröning also said that the structure of the camps had distanced him emotionally from the horrors he was part of. He described a healthy social life, with sports leagues and dances. His work—tallying and sorting money from prisoners—facilitated the machinery of death but kept him far from the physical act of murder. In other words, he convinced himself that because he only rarely saw the suffering he caused, he wasn't fully responsible for it, and certainly not in a legal sense. He'd enabled enormous human misery, but he hadn't directly, tangibly hurt anyone. In Gröning's mind, that made him innocent in the eyes of the law. I think most of us would disagree. The courts did. Gröning was found guilty and eventually sentenced to four years in prison.

Though the atrocities my father helped prosecute at Nuremberg were very different from the harm being done by oil and gas today, the ideas my father grappled with are just as relevant. The nature of complicity and responsibility—both our own and others'—is a moral issue, perhaps even a spiritual one. And it is as much a part of understanding and fighting climate change as it is understanding the warming effect of greenhouse gas.

This is especially the case when it comes to what I think of as the true believers. In 1980, or 1990, maybe you could have worked in the energy industry and been genuinely unaware of the terrible side effects of burning fossil fuels. But today, if you're the CEO of an oil and gas company or an industry lobbyist, that's just no longer possible. You're likely wealthy and well-respected by many of your peers. You're living a perfectly normal life. Yet you owe your wealth and social position to your willingness to knowingly inflict misery on millions, if not billions, of human beings.

Throughout history, people have found all sorts of ways to justify this kind of behavior. The old justification was to pretend climate change wasn't real, or to cast doubt on the science. But even that was a flimsy pretext. If you drink six beers and then get in your car and drive home, there's no guarantee you'll kill someone—but it's still wildly dangerous and irresponsible. For decades, the fossil fuel industry was driving drunk. Only there were billions of people in the car with them.

As the consequences of climate change became impossible to deny, people within the fossil fuel industry convinced themselves that the human suffering they cause is necessary. We need oil and gas, they argue, to enjoy modern life. Often, they go even further. "We're the adults in the room," they say, "because we're the ones willing to do what needs to be done." This kind of thinking might help people feel better about the choices they've made in life. But it's hard for me to understand the idea that it's immature to want to prevent a global humanitarian catastrophe.

What's more, the idea that oil and gas is a necessary evil simply isn't true. While we can't make the transition away from fossil fuels literally overnight, if the fossil fuel industry stopped using its political clout to block the transition to cleaner sources of energy, we could make the needed transition a lot faster than most people think. In fact, we could be most of the way there by the end of this decade.

Meanwhile, the value of an oil and gas company—and therefore the success of an executive at that company—isn't tied to the amount of energy they can sell today. It's tied to the amount they can sell for years, and even decades, to come. If you work for the fossil fuel industry, you should be clear-eyed about it: your job is to keep us addicted to fossil fuels for a very, very long time. That inevitably means bullshitting the rest of us on climate. Increasingly, it also means trying to throttle the better, cheaper, cleaner sources of energy that pose an existential threat to oil and gas.

The fact that fossil fuels are currently a "good" business, in the sense that the companies engaged in it make a lot of money, doesn't justify working in that business today any more than the fact that cigarettes were a "good" business in the 2000s justified someone's decision to work at Phillip Morris.

Part of the problem is that when it comes to oil and gas, the money is tangible, while the harm is not. The catastrophes that arise from extracting more oil from the earth might not occur tomorrow, and when they do, they might not be directly attributable to you. But just because you'll never see the harm you cause doesn't make it less real. Nor does the fact that we can't determine—yet—the exact contribution a new fossil fuel project makes to some future drought, wildfire, or other disaster. Just because burning oil and gas kills people in a less direct way than pulling the trigger on a gun doesn't absolve people of responsibility any more than it absolved pharmaceutical executives of responsibility for the opioid epidemic, or tobacco executives for the lung cancer their products cause. If you go to work every day to help burn more fossil fuel and dump more carbon into our atmosphere, you're devoting your life to the planet's death.

And I'm not only talking about the culpability of wealthy executives. If you're truly Jean Valjean—if the only way you can feed your family is by working in the fossil fuel industry—then I don't feel I'm in a position to judge. But the fact is that whether we're talking about money, power, or political influence, the people who benefit the most from the fossil fuel industry's success *do* have a choice. The advertising executives who come up with greenwashing campaigns; the accountants who help oil and gas companies secure huge tax breaks; the bankers who fund new drilling; the management consultants who help energy companies reorganize and restructure; the law firms that shield fossil fuel companies from liability; and so many others. It doesn't really matter whether you work in the fossil industry or merely with it. Either way, your success is tied to human suffering. There are better ways to spend one's time on earth.

That's even true for the person I consider to be the best investor in history, Warren Buffett. I would never say Warren and I are close, but every time I've spoken to him, or heard him describe his approach to his profession, I've come away admiring his discipline, foresight, and judgment. Which is why I'm concerned that not too long ago, Warren's company, Berkshire Hathaway, made big new investments totaling tens of billions of dollars in oil and gas. Of the 8 billion people alive today, it's possible that no one is better than Warren at predicting the future of commercial activity. We're talking about "the Oracle of Omaha." And he's predicting that decades from now, there will still be a large market for fossil fuels.

For the record, I think and hope that Warren is wrong. There are two reasons to invest in fossil fuels. One is that you think the world is going to transition away from oil and gas, and that the value of fossil fuel companies will eventually go down, but that you can sell before that happens. In investing, this is called "catching a falling knife" and it's almost never a good strategy. It's also not the kind of investing that Warren does. Instead, he tries to figure out what the future will look like and make investments based on those predictions. His reason for investing in fossil fuel companies is that he thinks over the long term they're going to do very well.

Warren has an incredible record of being correct and disproving his skeptics. He's also a good man. But in the case of fossil fuels, he doesn't seem to recognize how quickly our ability to develop and deploy clean energy is growing—and how cheap that energy is becoming. His firm also seems to have a blind spot where climate change is concerned. In 2015, his right-hand man, the late Charlie Munger, said, "I don't think it's totally clear what the effects of global warming will be on extremes of weather." Well, yeah. Nothing in the future is totally clear. What's 100 percent clear, though, is where we're heading—and if Berkshire Hathaway is still making its investment decisions based on the view that we need to know exactly how the climate will change before we take any meaningful action, that's going to hurt their decision-making.

Warren isn't anti–clean energy—the utility companies owned by his firm were some of the first to recognize the opportunity in switching to more renewables, and he deserves credit for that. But I worry that part of Warren's thesis is that fossil fuel companies have so effectively enmeshed themselves in certain parts of the American political system that long after they're not commercially viable, the government will prop them up. Here, too, I strongly hope—and believe—that he's mistaken. Right now, climate change isn't enough of a top-of-mind issue for voters, especially in red and purple states, to make legislators who do the fossil fuel industry's bidding worry about losing their jobs. But as the climate changes, and the impacts are felt more directly by more people around the country, attitudes and votes are changing, too.

Still, let's do something that people rarely lose money by doing: let's assume Warren Buffett is right. What would have to happen for his massive investment to generate the returns he's looking for? We'd have to continue to extract and burn enough carbon that our planet would be plunged into nonstop, essentially irreversible catastrophe. In other words, if Warren is wrong, he's wrong—and if he's right, society is screwed.

You're probably not deciding whether to invest tens of billions of dollars in oil and gas. But maybe you do something like I did when I helped Alaska invest its windfall tax revenues or facilitated the mergers of oil companies during my first jobs out of school, at Morgan Stanley and Goldman Sachs. Hurting the planet wasn't my full-time job (and to be fair, in the 1970s and 1980s, the science really was less clear and widely available than it is today). But at the same time, I was a cog in the machine that was profiting a handful of companies and people at the expense of the vast majority of us. While society considered my job to be pretty impressive, the "not my problem" approach is the wrong approach.

It's like the question, "What did you do in the war?" For people in the Greatest Generation, if the answer was, "Nothing," or even worse, "I worked for a company that didn't technically break the law but helped the Nazis fuel their war machine," that was deeply embarrassing, and rightly so. I think the same thing will be true ten or twenty years from now. If you don't do your part on climate change now, when you had the greatest chance to make the largest impact, you'll always regret it.

To put it in a more positive way: if you want to live a meaningful life in an era that will be defined by climate change, then fighting climate change needs to be a meaningful part of your life. There's no excuse for fiddling while the planet burns.

Nor can you throw up your hands and say, "This is all too scary and difficult for me to deal with." Which brings me to one group of fiddlers in particular: climate doomers.

To some extent, I'm talking about wealthy doomsday preppers, such as right-wing donor and tech mogul Peter Thiel. He hasn't done much to fight climate change. In fact, his tens of millions of dollars in political donations to climate deniers have made the problem much worse. But he's reportedly bought property in New Zealand apparently on the theory that climate change is unavoidable, and that if he relocates thousands of miles away, he can protect himself, his friends, and his family.

This is an extreme case, since most people can't afford to buy a climate refuge. What I'm more concerned about is the impulse to become despondent. I find that this attitude is particularly common among young people. Almost none of them doubt the existence of climate change, and many of them are leading the fight to stabilize the planet.

But some of them take for granted that runaway climate change will destroy life as we know it. In fact, a recent survey of 10,000 sixteen- to twenty-five-year-olds published in *The Lancet* found that more than half of them agreed with the statement "Humanity is doomed."

Young people are furious—and rightly so—at older generations for leaving them a huge mess. But to go beyond anger to despair plays perfectly into the fossil fuel industry's hands. It's also completely wrong. Sure, one day we might get to a point of no return, where climate change is happening so quickly, and the oil and gas industry has been so successful at keeping new technologies from emerging, that there's nothing we can do. But right now, we're not even close to that point.

In the end, climate doomers are just finding a different way to justify the decision to do nothing. Despondency is just another form of compliancy. Because none of the science suggests that we're doomed.

There's no guarantee that we're going to prevail. Stabilizing the planet will take intelligence, hard work, creativity, maybe even some luck. No, we don't know exactly how things will play out, even in a best-case scenario. But that doesn't mean we've lost. If we do everything we're capable of, I think our odds are actually really good.

It's uncomfortable to be the guy saying that the vast majority of us aren't doing enough to fight back. But it's the truth. Maybe you've started to take action, but you still think that getting to net zero is someone else's problem. Or maybe you just don't know what to do to address the crisis we're facing and you're letting the magnitude of the threat overwhelm you. You think about climate and worry about climate, but you haven't yet turned those thoughts and worries into meaningful action. If you're reading this and thinking, "I need to do more," there's no need to feel ashamed.

In fact, when I consider the gap between the number of people who care about climate and the number of true climate people out there, I don't feel discouraged. It's the opposite. If we were doing everything we possibly could as a country, a society, or a species, and the planet was still warming at its current alarming rate, *that* would be discouraging. But we're barely scratching the surface of what we can do. Our incomplete effort is already reshaping the way we create and use energy, revolutionizing transportation, making us rethink agriculture, developing technologies that used to be the stuff of science fiction, and so much more.

Imagine what would happen if more people stopped rooting for the end of the world. I'm not talking about all 8 billion people on the planet. Say, for the sake of argument, that just 10 percent of us devoted just 10 percent more of ourselves to climate. Those numbers aren't exactly measurable, of course, but you get the idea. The effect from that small change alone would be enormous.

I don't just view doing our part on climate change as a matter of science, or policy, or even maintaining a livable planet, as crucial as that is. I see a failure to do more on climate—our continued, collective insistence on rooting for the end of the world—as a cause for deep spiritual concern. I believe in God, but I'm not talking about religion here. I'm referring to the sense of spiritual crisis we feel, as individuals and as a society, when we try to ignore the single most existential threat we face. Imagine living in a house that's on fire yet trying to go about your business—watching TV, eating dinner, reading a book—as though nothing unusual were going on. Think about how profoundly strange that would be.

No wonder our country feels adrift. No wonder so many people do, too. Right now, most of us are living in either a state of active denial or profound contradiction. We're not doing our best, and that leaves us feeling empty. Imagine how much better off our lives would be if we stopped rooting for the end of the world and decided that business as usual wasn't good enough anymore. Climate change is here, and while no one is happy about its arrival, the fight to stabilize our planet can give our country a sense of direction it's lacked for far too long. This is America's opportunity to do what our country does best, what the Greatest Generation did all those years ago. We can lead the world. We can save the world. And we can make a ton of money in the process.

What is true for America is also true for individual Americans. Three years ago, my big brother Hume told me, "I'm looking at this climate stuff, and I'm so depressed. I don't know how to think about it. It's just getting me down." So, I said, "Let me give you some advice: *do something.* You don't have to know every step. You just have to take the first step. Figure out how you can get in the game. When you do, you'll find that, actually, it's really relaxing."

That's exactly what happened. In the years since, Hume has gotten more and more involved with environmental organizations, especially the Open Space Institute, where he's a board member. I've seen how much he enjoys the work, and how meaningful it is to him. Not coincidentally, he's been a lot more positive and a lot less depressed about climate.

Like most people, I can also get stressed about what the science is telling us, or the ways that our planet is already beginning to change. But I don't stay stressed. Becoming a climate person isn't always easy, but the alternative—shirking my responsibility to help solve the biggest global problem of our time—would be far more stressful and upsetting.

Trust me: when you stop rooting for the end of the world, it doesn't mean giving up the things you enjoy or that get you out of bed in the morning. It certainly doesn't mean ignoring other important issues. It means harnessing your talent and drive to an issue that, if left unchecked, is going to make nearly everything else seem inconsequential by comparison.

The years I've spent as a full-time climate activist haven't been the easiest of my life. I could have stayed at my firm, not caused trouble, and made a lot more money. But I would have missed out on the most fulfilling professional experience I've ever had—and with the exception of my family, the most fulfilling personal experience I've had, too. There's an inner peace that comes with knowing you're doing your part. I wouldn't give that up for anything.

In the end, maybe that's the best reason of all to become a climate person. Saving the planet isn't simple. It's not guaranteed to happen. But it's a really good reason to walk around planet Earth on two legs.

CLIMATE PEOPLE

Priya Donti

From helping doctors diagnose stroke patients to improving weather predictions to synthesizing the world's information in programs like ChatGPT, artificial intelligence and machine learning have already changed our day-to-day lives. Priya Donti wants to make sure AI changes how we solve climate, too.

Priya became interested in climate as a high school freshman when her biology teacher spent the first week of class on sustainability. Over the next few years, she did what she could to reduce her carbon footprint and volunteered with her college's environmental club while majoring in computer science. Then, around the time she was graduating, Priya read about the ways AI could help power grids use more renewable energy, and she suddenly saw a way to combine her interest in climate with her interest in computers.

On a post-graduation fellowship, Priya spent a year interviewing nearly 150 people in Germany, India, South Korea, Japan, and Chile about AI-influenced power grids, then worked on developing AI algorithms for them for her PhD. After presenting her work at a 2018 machine-learning conference, she was invited to contribute to a paper on AI's potential climate impact. The paper was supposed to be just an overview of how the technology could be used, but after other computer scientists started asking how they, too, could get involved on climate, its authors got together and started Climate Change AI, a nonprofit that connects people from tech and climate backgrounds to help develop AI climate solutions.

The group's work includes mentorship programs that match those looking to get involved with people who have complementary expertise, grant programs to help develop promising technologies, and venues to discuss AI breakthroughs and how they're being used. As Climate Change AI's cofounder and chair, Priya works with the private sector, government, and other nonprofits to facilitate new climate solutions—and this work is already making an impact.

For example, a project called Open Climate Fix pursues the challenge that first drew Priya to climate—using machine learning to optimize renewable energy usage. On cloudy days, power grids that typically rely on large amounts of solar energy often turn to their reserves, typically oil, gas, or coal, since the utilities would rather not run the risk of running low on power. Because they don't feel they can accurately predict how much sun there will be, they play it safe. But this means that solar energy generated on sunny days often goes unused. Using AI to analyze years of weather trends, Open Climate Fix worked with the UK's national power grid operator to develop a system that cuts the error of solar energy generation predictions in half, which can limit the need to tap into oil and gas–based reserves.

Halfway across the world, meanwhile, scientists backed by a Climate Change AI grant are using machine learning to help counteract flooding in Fiji. By examining satellite images and getting information to government officials, this technology is helping improve local disaster responses, protect crops, and support the country's farmers.

Notably, the programs incubated at Climate Change AI are all the results of early-stage technology. More advanced AI systems will be able to help make buildings more energy-efficient, eliminate water waste in agriculture, and much more.

"There's no one entity or person in this sphere that is going to be able to do this alone," Priya says of AI's climate role. But that's why Climate Change AI exists—and why its mission is so important. "The ways in which different organizations can boost each other up and bolster each other's work is going to be really critical to tackling this."

CHAPTER V

THINK LIKE A WALK-ON

have a friend who I think of as the ultimate team player. He's been part of many institutions and organizations on Wall Street, in government, and in Republican Party politics. He expresses his opinion *within* those club-like structures. But he never, ever turns his back on his team.

For much of his career, that's been smart. Frankly, it sometimes seems like he knows something I don't. His attitude has helped him succeed in both finance and government, and he's been able to make a lot of positive change.

At one time, that included climate activism. In the early 2010s, he and I, along with a bunch of other businesspeople, worked on a project to bring private-sector leaders together to make the economic case for preventing the worst effects of global warming. Almost immediately, our project started gaining traction, in part because we were represented by people from across the political spectrum who had credibility in the corporate world. No one could accuse people like me of being socialists. They *really* couldn't accuse people like my friend of being socialists. The private sector began to pay more attention to the risks climate posed to their businesses. It seemed like we could really accomplish something big together.

Then, in 2016, Donald Trump became the presumptive nominee of the Republican Party, and my friend immediately pulled out of our project. He knew climate change was real. He understood exactly how much harm it would bring to the country, to the world, to his grandchildren. He cared about stopping it. But he also knew that Trump—who in 2012 tweeted that "The concept of global warming was created by and for the Chinese"—was going to echo the most extreme, conspiracy-minded elements of his party. He felt that if he angered the Trump campaign, his influence within the GOP would disappear. Better, he thought, to stay in the game and not rock the boat.

"How can you think that way?" I asked. "What we're saying is true. It's important."

He said, "Yeah, but I'm a Republican, and if I don't have Republican credentials then people around the world who I care about, and who I do business with, won't give me any credibility and I'll lose the ability to make *any* impact. I can't afford to piss off Trump." He wasn't willing to tell the truth about climate, because doing so would mean risking his importance, influence, and access.

In ordinary times, there might be wisdom to this approach. Staying on the inside usually means you have to work incrementally, but sometimes incremental progress is the best you can do. Also, if you abandon the playing field entirely, you risk losing your ability to be heard. Other people might take your place who don't share your values or possess your judgment.

But this isn't an ordinary moment for our planet. We can't afford to ignore the truth in order to work within the system, but at the same time, we can't get much done if we only work from the outside. If we're going to stabilize the planet, what we need is a middle ground, a position that allows us to work within established institutions and ways of thinking, but that also allows us to step outside of them and challenge them when necessary.

One of the defining experiences in my life took place in just such a middle ground. I wouldn't have predicted it back then, but the mentality I learned as a walk-on soccer player has been hugely helpful throughout my life—and it might even be essential for climate people today.

When I arrived on Yale's campus in the fall of 1975, joining the team had been easy. That summer I'd received a letter from the head coach, Bill Killen, inviting me to try out. (It came with a workout plan so intense I genuinely thought it was a joke. I ignored it and prepared for the season by playing tons of pickup soccer in Central Park.)

I was also confident—maybe a little too confident—in my athletic abilities. I was a three-sport athlete in high school, and I didn't see why I couldn't do the same in college. It seems insane now, but my plan was to play varsity soccer, varsity basketball, and varsity tennis. I knew that some athletes had been recruited by the school, while I had only been invited to walk on to the team. But it was only after I got to the first practice that I realized that walk-ons were in their own, lesser category. The recruits were treated differently. The coaches knew their names, were interested in their studies and what dorms they were in, and wanted to make sure they were doing okay away from home. I got none of that special treatment, but I figured it wouldn't make much difference once we got on the field.

I couldn't have been more wrong. Our coaches had recruited some amazing athletes, but as strange as it sounds, some of them weren't great soccer players. One of the recruits was blazing fast but didn't know how to dribble the ball with his head up. He kept looking at his feet. Another guy could shoot the ball like a rocket and was even athletic enough to do a flip in celebration each time he scored a goal. But that didn't happen often, because he took too long to get his shots off. Before he could slam the ball into the net, a defender would run up and steal it from him. To the coaches, however, it didn't matter: the recruits were the recruits, and the walk-ons were just walk-ons.

I was pretty sure I was better than some of the varsity players, but as a freshman, I was assigned to the JV. So were a lot of other really good soccer players. We knew we had talent, and we loved being a team—some of my best friends to this day are people I met playing JV soccer. All season long, we backed up our confidence with results. We played against all the other Ivy League schools and a bunch of others and went 12–0. But what we really looked forward to was our intramural scrimmages against the varsity team. We played them twice a week, every week, all season long, and we beat them every single time.

In my teenage naivete, I thought that when JV walk-ons consistently beat the varsity recruits, at least some of the JV squad would get to move up. Instead, Coach Killen told us, "You've got to stop winning, because you're sapping the varsity's confidence."

This is where I first truly experienced what I think of as the walk-on mentality. I've always loved being part of teams. I wanted to be coached. But when I heard Bill Killen tell us to start losing, my first thought was, *no way.* He chose subpar players, he lost, and now he wanted *us* to stop trying? Not gonna happen. I never thought about quitting the team. But I also never thought about playing along. It was the most ridiculous thing I'd ever heard.

I soon realized that I wasn't making varsity my freshman year under any circumstance. It didn't matter if I deserved it or not. The coach was never going to admit to himself that he'd recruited the wrong players. He wasn't really committed to them, but he was committed to his previous judgment.

While I was disappointed, there was a silver lining. I loved playing soccer with my JV friends. In fact, we banded together and made a pact: we would only go up to varsity if they agreed to give us real playing time. My sophomore year, Coach Killen put the pact to the test: he offered me a varsity spot but told me I'd be warming the bench, backing up one of the recruits. A pure team player might have given up on the pact and taken the offer. I stayed on JV.

I finally did join the varsity team as a junior and was captain my senior year. But looking back, I wouldn't have traded my walk-on experience for the world. If you offered me the chance to time travel, arrive on campus as a recruit, and play every minute of every varsity game for all four years, I would turn you down. Because I learned something over those four years—quite possibly the most important thing I learned in college. You can't go it alone in life. But you also can't just go with the flow. The higher the stakes, the more important it is to be a team player *and* be independent at the same time. That might sound like a contradiction, but it's not.

Today, in the climate world, we need independent team players more than ever. On one hand, our world needs to change quickly—and we need teams to make that happen. We need new technologies, new ways of creating energy, new ways to cut emissions of greenhouse gas. Then, we need to figure out how to deploy these new technologies worldwide, in countries with wildly different levels of infrastructure, prosperity, and political stability, all while facing relentless opposition from fossil fuels. Only by working together can we achieve the kind of change we need at the pace we need. That's particularly true within institutions and organizations—everything from media outlets to businesses to governments to banks. If people who already have credibility within existing institutions can help those institutions transform to meet this threat, it's going to make an enormous difference for the climate and the planet.

On the other hand, organizations can become havens for groupthink. They can be corrupted by special interests that don't care about the future of the planet we all live on. (My friend's Republican Party, under Trump, is an excellent example.) To be a climate person, you have to be able to play the outside and inside games simultaneously—to work within big organizations to drive change, and then, when necessary, to challenge those organizations to change themselves.

Above all, you can't fall victim to the idea that what's good for the team is necessarily good for the world. Institutions and organizations are a means to an end. To harm the entire planet in order to further the interests of a single business, political party, or nonprofit organization is a much more high-stakes version of starting the recruits because they're the recruits, even when it makes it harder to win.

Striking this balance is easier said than done, particularly when the issues are complicated and the potential consequences so enormous. It can be especially difficult when the people dispensing conventional wisdom are respected and important. But in my experience, part of thinking like a walk-on is remaining skeptical of those I call "suits with titles."

I've always been amazed at how many people are willing to respect someone just because of the position they hold. A few years after I went to Morgan Stanley, while I was still on Wall Street, I was talking with a friend about a wealthy corporate raider. I thought he was making a really stupid investment decision and said so. "He's got a lot more money than you," my friend reminded me, "so he's probably a lot smarter than you are."

I said, "Really?" My grandfather wasn't rich, I told my friend—but he was smart as hell. If someone had ever dared to suggest to him that people with more money were automatically smarter than he was, I don't think he would even have gotten angry. He just would have looked at them like they were crazy. (It turned out I was right. Over the years, as I made a lot of money, it didn't make me any smarter. It just meant I had more money.)

In the climate world, there are plenty of suits with titles who help prop up the fossil fuel industry: CEOs and politicians and investment bankers and heads of industry-funded think tanks, all of whom are used to being treated as people with serious ideas because they hold important positions.

If we're being honest, the climate movement has its own version of this kind of problem, where people end up reinforcing conventional wisdom not because the arguments are the strongest, but simply because the institutions who espouse them are important. The mainstream climate groups—the well-funded environmental organizations that have been around for decades —have racked up extraordinary achievements over the years: they defended and strengthened the Clean Air Act and Clean Water Act; curbed acid rain; protected the Arctic National Wildlife Refuge; won the Supreme Court case that allows greenhouse gas to be regulated as a pollutant; and much more.

Our society is much better off because of these groups, and I've been proud to support them over the years. However, in part because their origins are in conservation and opposing development, they have sometimes been too incremental in their advocacy on climate change or let the perfect be the enemy of the good.

A good example involves permitting reform. Last year, a bill before Congress would have sped up the government permitting process for new clean-energy projects. That's absolutely essential if we're going to meet our emissions-cutting targets. But many of the most influential conservation groups opposed the bill because it would also have sped up the permitting process for new natural gas pipelines, and because even clean-energy projects often come with environmental tradeoffs. (For example, placing new wind turbines on an undeveloped mountainside, or building a solar farm that could potentially disrupt the habitat of a threatened species.)

Hamstringing our ability to fight climate change because it would come with some conservation-related tradeoffs is terrible for the environment in the long term—and just because some of the organizations adopting this strategy have done truly impressive things in the past doesn't make it a good strategy. Bill McKibben wrote about this eloquently in *Mother Jones*. Describing a different set of clean-energy-related tradeoffs—that lithium mining has the potential to hurt ecosystems and many communities in Central and South America—he acknowledges that these harms are real, unfair, and tragic. But, he writes, reflexively saying "no" to new development ignores the harms of maintaining the status quo:

> [S]lowing down lithium mining likely means extending the years we keep on mining coal, that more than 6 million people a year die from the effects of breathing the byproducts of fossil fuel combustion, and that we're dancing on the edge of the sixth great planetary extinction.

He went on:

> Making the perfect the enemy of the good is, in such a case, more like making the perfect the enemy of anything at all. When you're in an emergency, acting at least gives you a chance; not acting guarantees an outcome, and not a good one.

In my experience, it's often the people outside the climate establishment who have been quickest to understand the changing nature of the threats we face, and how to fight them. Part of thinking like a walk-on is having the courage of your convictions and not outsourcing your judgment to others just because society has named them "coach" or "senator" or "CEO."

There's another version of ignoring suits with titles: great leaders and managers, in my experience, make it clear that they care more about the idea than the title of the person who has the idea. This was certainly true of the best boss I ever had, Bob Rubin.

I met Bob when I started working at Goldman Sachs. At the time, he was already well-known as the young head of the firm's highly profitable risk-arbitrage division and as a prolific fundraiser in Democratic political and economic-policy circles.

In a lot of ways, Bob and I are very different. He likes to consider all his options; I can be more impulsive. With the exception of his time in Washington, he's spent his entire adult life in New York City, while I still look back on our move to California as one of the best decisions I've ever made. Politically, he's closer to the center on many issues than I am, although I give him real credit for coming around on climate.

But while we don't always take the same approach to issues, from the moment I began working for him, I was struck by Bob's judgment. When I got to Goldman Sachs, one of the firm's biggest clients was a guy named Ivan Boesky. One day, Bob told us, "Ivan Boesky is doing something wrong, and he's going to get in trouble. We can't be connected with that. We're not going to forbid him from working with Goldman Sachs—you can take his order if you're a trader. But no one is to ever speak with him or interact with him, not even on the phone." At the time, I didn't think that much about it. But two years later, when Boesky was arrested and revealed to be the ringleader of a giant insider-trader scheme, I realized that Bob had an ability to see two moves ahead.

And that was only part of what made Bob Rubin such an exceptional manager. He was also a good listener; he was clear about what he needed you to do; and he kept your focus where it belonged.

Perhaps most important of all, as long as you were honest, took responsibility for your actions, and ran a thorough process, Bob treated you with respect and listened to what you had to say. If you were a junior person, and he thought you had the best argument on an issue, he'd take your side. Bob Rubin is the quintessential establishment figure, yet I think he has just a little bit of walk-on in his personality: he makes his decisions based on what people have to say, not on who says it.

In fact, the one time I remember Bob speaking harshly with me was when I let unnecessary, arbitrary bureaucracy get in my way. It was

Christmas Eve, 1984, and we had both gone into the office. Things were quiet. Then, out of nowhere, we got some information that completely blew up a position that the arbitrage department had taken in Phillips Petroleum, which was fighting a takeover bid from oilman T. Boone Pickens. Bob immediately said, "Tom, do the following five things . . ."

"Bob," I explained, "I'm not working on that deal."

"This is a problem we need to handle right now," he snapped. "Do what I tell you."

And you know what? He was right. The firm had a major problem, and we were the only two people in the office. Who cared that Phillips Petroleum wasn't technically my responsibility? Assignments are important for teams to function, but they can't be more important than the team's actual mission. I should have just gotten to work.

It's a lesson I've tried to carry with me ever since. You've probably heard the phrase "Stay in your lane." In my experience, the walk-on mentality is about doing the opposite—getting out of your lane and focusing on accomplishing what needs to be done rather than saying, "It's not my job."

It's probably no surprise by now that getting out of my lane has rarely been a problem for me. I've certainly never thought of myself as an establishment person. I worked for decades in investing, but I never felt like a Wall Street insider. I supported campaigns for major ballot measures, formed a nationwide political organization focused on turning out young voters, and even ran for president, but I've never felt like a political insider. I've dedicated years of my life to climate, and I work with brilliant scientists, entrepreneurs, activists, and policymakers every day, but I don't think of myself as a climate insider, either.

Being not-quite-on-the-inside has given me important perspective. But I don't mean to imply that it's easy. In fact, not being an insider can be really frustrating. Thinking like a walk-on can also come with real personal and professional costs. In 2008, after Barack Obama's election, some people encouraged me to consider trying to join his administration. The Great Recession was in full swing, so leaving my team would have been basically impossible, but I was still kind of intrigued. Washington is a heard-it-through-the-grapevine kind of place, and I began to hear that my name was in the mix for some fairly high-level positions. It started to seem as if I might get a call from the president.

I didn't. Later, I found out at least one reason why. Someone I know well, and still consider a friend, had given White House officials their honest opinion of me: whatever my other good qualities were, I wasn't a team player.

Was I annoyed when I found all this out? Absolutely. Not least because I've played, and won, on plenty of teams.

But in retrospect, maybe my friend had a point. Maybe I'm not cut out to be staff in an administration. In those jobs, you don't have to go along with everything, but you do have to be willing to hold your nose at times. I don't know if I'd be comfortable doing that. My insistence on maintaining independence has probably cost me a lot of other opportunities through the years as well. It's definitely stressed me out at times. My brother Jim, who's an idealist but also a pragmatist, is always saying to me, "Tom, you get so upset—just go along with it for Christ's sake."

But I guess I'm just not a go-along type of guy. That's the cost of not being an insider: a kind of frustration tax.

Climate people have to be prepared to pay the frustration tax in order to protect our planet. Many of the people I trust most on climate were, for years, ridiculed as alarmists for making predictions that turned out to be true. When it comes to the facts about the speed at which our planet is warming, and what that warming is doing to our world, being a climate person often means being a bearer of bad news. Even among people who understand the issue, and broadly agree that action is needed, climate people are used to being the fly in the ointment, telling those around them, "This is great, but we've got to do more."

In 2015, NextGen Climate Action, an organization I launched to help mobilize young people to fight climate change, began our "50by30 Campaign," calling on candidates to commit to generating 50 percent of America's energy through renewables by 2030, and 100 percent by 2050. While some Democratic candidates were talking about climate, we didn't think anyone was talking about it enough. We saw the campaign as a way to generate interest in an issue that would help define America's future while holding the most powerful people in the country accountable.

Democratic Party officials were furious. Even progressive candidates refused to agree or even respond. The consensus was that we were calling for too much, too fast, and harming Democrats' electoral prospects, making it less likely we'd get anyone to act on climate. The conversations I had arguing our side were not fun.

But you know what? We were right—and not only on the science. I think we were right on the politics, too. Within months, both Hillary Clinton and Bernie Sanders had signed our renewable energy pledge. By the end of 2016, New York and California had both adopted 50by30 standards. Fast forward to 2020, when, as the Democratic nominee for president, Joe Biden called for 100 percent clean energy by 2035—a goal significantly more ambitious than that laid out by our original 50by30 pledge from just five years earlier, and one that our polling showed made a big difference for him among younger voters.

Today, sixteen states—California, New York, Colorado, Hawaii, Illinois, Maine, Maryland, Minnesota, Nevada, New Jersey, New Mexico, Oregon, Rhode Island, Vermont, Virginia, and Washington— have adopted 50by30 or better standards. Through the Bipartisan Infrastructure Law and the Inflation Reduction Act, Biden has been able to deliver on far more of his climate promises than most pundits would have predicted after his election. And being strong on climate is now almost a prerequisite for Democrats running for office.

Paying the frustration tax is no fun. And you won't always get credit, so you have to find your rewards in reaching your goals while doing what you think is right. That's not just true for individuals; it's true for whole societies, too. At its best, America is willing to pay the frustration tax—to lead the world while at the same time trying to live up to our ideals and accepting all the difficulties and challenges that come with that mission.

In fact, one of the things I love about America is that we're a walk-on country. No, we definitely aren't a true meritocracy. As the former Dallas Cowboys coach Barry Switzer famously said, some people are born on third base and go through life thinking they hit a triple. But if you look at our history, for all our considerable, often tragic shortcomings, we're the country that rejected the hierarchies of aristocracy, monarchy, class, and caste. At heart, for all the challenges we face, and for all the work there is ahead of us, we're still that country. You don't have to be descended from the Mayflower or come from money to make it in America. Over 43 percent of America's Fortune 500 companies were founded by immigrants or their children. That's incredible. We don't always live up to our ideals, but at our best, we're the country that believes everyone should have a chance to reach their full potential.

When it comes to walk-ons, and to climate people, no one better illustrates this than Eddie Garcia.

Eduardo Garcia was born and raised in the Coachella Valley, a place that, like so much of our country, is home to the extremes of haves and have-nots. Eduardo comes from a family of have-nots. He went to the local public school and then enrolled part-time in a community college while he worked. It took him five years to graduate. But from there, he got his bachelor's degree at UC Riverside, completed the public administration program at the Kennedy School of Government at Harvard, and got his master's in public policy from USC. He was elected to the Coachella City Council, then mayor, and then, in 2014, assemblymember for the community where he grew up—in one of the poorest districts in the state.

I have a friend who belongs to a fancy country club near Palm Springs, California. At one time, at least one of the Koch Brothers was a member there—it's that kind of place. Until the lines of his district were recently redrawn, Eddie represented Palm Springs, and long after I'd gotten to know him, I said, "Hey, Eddie, do you know this country club? It's got golf courses and all that stuff." He goes, "Oh yeah, I know it."

"No," I said, "I mean did you know that it's in your district?"

"Tom," he told me, "I know. My mother cleaned the houses in that country club."

In other words, Assemblymember Garcia never got to decide whether to be an outsider to the power structures of business and politics. He was born outside them. But I don't think it's a coincidence that he's been one of the best lawmakers on climate in the state. He works harder than just about any public officeholder I've ever met. If he believes something needs to happen, he'll be on the phone at all hours of the night, calling everyone he knows until it gets done. His first year in Sacramento, he became the only freshman legislator ever to have twelve bills signed into law.

Eddie is the consummate walk-on. He's dealt in the difficult, trade-off-filled, art-of-the-possible world of politics, but he's always fought for climate action, even when it would have been so much easier not to. He's never sold the movement out. In 2016, he helped lead the charge to update the California Global Warming Solutions Act and mandate that we get our emissions to 40 percent below 1990 levels by 2030.

Eddie's the type of person too many environmental and climate organizations have overlooked for too long. But he doesn't care. He knows how much climate change will hurt districts like his, he knows that California leading the way on climate will mean less pollution and better jobs for his constituents, and he knows that as great as it is to be on a team, you can't forget why you signed up in the first place. He gets things done for his community, even when it means going out on a limb to do what he thinks is right.

The type of work Eddie has done to protect communities like his from pollution—and make sure they have access to the economic opportunities of the transition to clean energy—is often called "environmental justice," which can sound as though it's a separate priority from fighting climate change. But from what I've seen, and more important, from what people who come from underserved communities like Eddie's tell me all the time—climate and justice are inseparable. We're all being harmed by burning oil and gas, but poor people in disadvantaged neighborhoods are being harmed the most. They're breathing toxic air from refineries. They're drinking toxic water from fracking. They're picking fruit and vegetables in ever-more-frequent heat waves. They're experiencing not just all the effects of extreme weather, but a health and safety crisis to go with it.

If we want to build a worldwide movement that will stand up to the fossil fuel companies and take meaningful action on climate, then we can't treat the people most affected by a warming planet as an afterthought. As George Shultz used to say, "If you want me in at the landing, ask me in on the takeoff." We can't muster political will tomorrow without including the people most harmed today. Whether it's the Civil Rights Movement that defeated segregation, the farmworkers who organized under Cesar Chavez, or the labor movement that won better wages and working conditions in the wake of the industrial revolution, real change has never come from the top down. That same kind of grassroots movement is helping to drive action on climate today.

From a climate policy standpoint, environmental justice is both critical to getting things done, and to getting things done right. That's something people like Eddie understand in their bones. And while Eddie is exceptional, he's not alone. In my experience, a disproportionate number of the most effective fighters for climate in our state have been people who represent lower-income, predominately Latino districts. Climate isn't ideological or theoretical for them. They know how much they have to lose, or to gain, depending on the choices they make.

You might not have the chance, as Eddie did, to pass bills that affect the world's fifth-largest economy. But no matter who you are, becoming a climate person is going to require you to balance being part of great teams with doing things the way you know is right—even if that means stepping out of your lane. If you're a filmmaker and people tell you, "Climate stories don't sell," figure out how to make one anyway and surprise them. If you're an investment banker, and you'd get more respect on Wall Street investing in a big bank that funds oil and gas than a smaller independent one that focuses on clean energy, stop worrying about what other people think and bring your talents where they're needed. If you have a platform—whether it's millions of followers on social media or just people at your dinner table—don't shy away from calling attention to climate, or organizing in your community, just because some people try to dismiss it as "political."

You don't have to be rude. But you do have to speak the truth—even though that means people will sometimes see you as rude.

The simple fact is that the establishments of the last fifty years aren't working, and when they are working, they're not moving fast enough. So forge your own way. Ignore the suits with titles, step out of your lane, and fight through the frustration to do what you know is right.

CLIMATE PEOPLE

Jigar Shah

To stabilize our planet, we'll need to develop amazing new technology. But that's only the first part. We'll also have to scale that technology so that it's adopted by as many people as possible as quickly as possible. Today, even as tech advances at a pace that would have seemed unimaginable just a few years ago, one of the biggest hurdles to building a clean energy economy is the cost of bringing new products and services to market.

Jigar Shah has spent the last twenty years figuring out how to solve this problem.

Growing up in Illinois, Jigar came across a book about the ways electricity is generated and was fascinated by it, specifically the parts about nuclear and solar. As a mechanical engineering major in college, nuclear seemed like the past and solar the future, so Jigar went with solar, landing an internship at an energy plant that manufactured solar panels.

In those days, anyone using solar energy had to buy and install the panels themselves—which was far too expensive, time-consuming, and complicated for most people. Why, Jigar wondered, couldn't solar work like other power sources? No one getting their electricity from fossil fuel had to burn their own coal or install a gas-fired plant in their front yard. Instead of pursuing an engineering career, he got an MBA and started a company, SunEdison, that bought and installed solar panels and then sold the electricity those panels generated to customers on a pay-as-you-go basis. His model worked, saving people money while helping the planet.

But Jigar wasn't done. After selling that company, he spent nearly a decade at Generate Capital, working to connect other clean energy and cleantech companies with the financing they needed to grow and helping to make technology like heat pumps, LED lighting, and electric buses viable on a large scale.

In 2021, President Biden and Secretary of Energy Jennifer Granholm recruited Jigar to run the Loan Programs Office (LPO) at the Department of Energy. LPO gives loans to companies developing promising clean energy projects, and the office got nearly $400 billion in new funding from the Inflation Reduction Act. It's their job to make sure this money gets to the companies that need it.

"We've called every one of those companies that have been labeled climate tech, whether it's green chemicals, green cement, green steel. It doesn't matter who it is, we've called them and said, 'Hey, let me introduce you to the loan programs office, so now we can help," Jigar has said of his role.

The first LPO loan after the IRA was passed went to a Utah facility that takes unused renewable energy generated from sources like solar panels and wind turbines and uses it to create green hydrogen, an alternative fuel that could one day power everything from cars to airplanes to spaceships. This green hydrogen facility is replacing a coal plant, boosting the local economy while becoming the first of many "hydrogen hubs" that will help power the western United States with renewable energy. Another LPO loan went to Ultium Cells, a company jointly owned by General Motors and LG Energy Solution, which develops cheaper, sleeker EV batteries. Their smaller size will allow them to fit in more models of cars and give engineers more leeway to design the EVs they power, helping to create new EVs that will look better than traditional cars, cost less, and could reduce gasoline use by up to 480 million gallons every year.

As Jigar told one interviewer, "It's great to meet the nation's best entrepreneurs and innovators and help them realize their dreams, which realizes American power." With the help of Jigar and his team, some of the country's most promising new companies are growing as fast as they can—and as fast as we need them to. They're not just helping us stabilize our planet and jump-starting the clean energy economy of the future, they're building a new generation of American jobs and ensuring that we remain at the cutting edge of the global economy well into the twenty-first century. No less important, they're saving consumers money, proving that when it comes to energy, the true premium being paid is the one being charged by fossil fuel companies.

"There's a lot of people who just are like, 'Can we do big things again?'" Jigar says. "And I hope to prove to people that we can."

CHAPTER VI

REDEFINE SMART

Barack Obama once called Jamie Dimon, the CEO of JPMorgan, "one of the smartest bankers we've got." He's not alone in thinking that way. I can't tell you how many times I've heard people—Democrats and Republicans, in finance and in media—say something similar.

To give credit where it's due, Jamie Dimon is extraordinarily good at what he does. He's built what is widely regarded as the best bank in the world. He's shown up and helped solve financial crises multiple times, working with the Treasury Department and the Federal Reserve. He's never been associated with any kind of cheating or illegality. There's a very strong case to be made that he's the world's best banker.

But there's more to the world than building a bank. And not too long ago, on a call with some of his wealthiest bankers, "one of the smartest bankers we've got" asked this: "Why can't we get it through our thick skulls that if you want to solve climate, it is not against climate for America to boost more oil and gas?"

It's easy to see why the head of JPMorgan would want that to be true. His bank has been, by far, the largest lender to oil and gas projects in the world. According to the Fossil Fuel Funding report for 2023, JPMorgan between 2016 and 2022 financed $434 billion worth of fossil fuel projects, more than $100 billion more than the next-largest lender. It depends on how you look at the numbers, but there's a very good argument to be made that JPMorgan is the world's largest fossil fuel company.

Very recently, JPMorgan has made some commitments to stop funding the most egregious types of fossil fuel projects, such as drilling in the Arctic and bringing new coal-fired power plants online. That's good news. But it will only make a difference around the margins. It's a drop in the bucket compared to the nearly half a trillion dollars of fossil-fuel lending overall. Even without new drilling in the Arctic, JPMorgan remains especially active in "frontier exploration," a fancy way of saying drilling in places remote and expensive to drill in. Those projects won't be online for at least several years and sometimes not more than a decade. Then, they have to sell the oil at a profit, which isn't easy to do in the short term because right now, oil from countries like Saudi Arabia is much easier to drill and therefore much cheaper.

So, when Jamie Dimon said "boost more oil and gas," he wasn't just talking about the next few years before we transition to something else. He was talking about the long term, too. For JPMorgan's hundreds of billions of dollars in investments in oil and gas to come good, demand for fossil fuels has to stay high for decades, which would mean pumping greenhouse gas into the atmosphere at completely unsustainable levels for decades.

We can stabilize the planet *or* we can fire up new frontier fossil fuel projects and lend extraordinary amounts of money to help the oil and gas industry grow. We can't do both. Jamie Dimon may be the world's best banker—but his comments and his bank's actions are a classic example of rooting for the end of the world. And I just don't think that's smart.

As climate people, we have to redefine "smart."

When I think about what it means to be smart, I think about the biblical parable of the talents. In the story, from the Book of Matthew, a rich man is about to leave home for a long time, and before he goes, he gives each of his three servants some talents. (In ancient Greek, "talent" was a word used to signify an amount of gold, sort of like "karats" in English.) The first servant gets five, the second two, and the third just one.

The first two servants invest their talents, doubling them. But the third one gets scared and buries his talent in the earth. When the rich man returns home, he's furious with the third servant. In fact, he casts him "into the outer darkness. In that place there will be weeping and gnashing of teeth."

That might be just a touch dramatic, but it's also pretty remarkable— the entire point of this more than 2,000-year-old story that gave us the word "talent" is that true worth isn't measured by how much you have. It's measured by *what you do* with what you have.

History is littered with examples of people who squandered their talents. Take, for example, two of the twentieth century's most talented filmmakers, D. W. Griffith and Leni Riefenstahl. Griffith basically invented the close-up, and Riefenstahl helped create the visual style still in use for everything from concert films to sports coverage. But what did they do with their talents? In 1915, Griffith created *The Birth of a Nation*, a violently racist movie celebrating the Ku Klux Klan. In the 1930s, Riefenstahl made propaganda films for the Nazis. Their films were technically brilliant, with groundbreaking technology and storytelling— all in the service of glorifying the KKK and Hitler.

Remarkably, as long as the films themselves were technically brilliant, a lot of people seemed not to care that they made the world worse. *Triumph of the Will*, Riefenstahl's most famous piece of Nazi propaganda, won the international prize at the 1935 Venice Biennale and the Grand Prix at the 1937 World Exhibition. In 2014, *Sight and Sound* magazine ranked it nineteenth in a poll of the greatest documentaries of all time. Meanwhile, in 1975, almost thirty years after his death, New York's Museum of Modern Art held a major commemorative showing of Griffith's work. A *New York Times* headline about the exhibit called him "The Man Who Invented—And Transcended—Film Technique."

To which I would say: who cares? I'm not denying that Griffiths and Riefenstahl were talented. But the defining fact of their lives is that they used their talents to do horrible things. If you make movies for Hitler, it shouldn't matter how good your movies are. But that's the talent trap: we tend to believe that talent alone makes you smart.

In fact, the talent trap might be even more dangerous today than it was back then. If you've got a lot of skill and ability, the world is full of opportunities for you to make money, and even earn some people's respect, doing really bad things. That's certainly true in fossil fuels. After all, keeping a dying industry alive for as long as possible—even when it's so obviously harming people—takes a huge number of talented people all working their hardest. CEOs, politicians, lawyers, accountants, consultants, scientists, marketers, media personalities, and more, all devoting themselves to promoting horse-and-buggy energy and delaying the switch to something better.

Here's just one example. In the 1980s, Americans began to worry about plastic pollution. In response, the plastics industry—a coalition that includes oil and gas, since plastics are made from fossil fuels—pulled off one of the most effective greenwashing campaigns in American history. The companies knew that recycling plastic doesn't work. It's too expensive, especially when you have to sort the plastic by type. Despite that, they lobbied state governments to mandate plastic recycling. They even added those tiny numbers you still see on plastic bottles today so that they could promote their new "coding system" as an alternative to regulation.

The messaging campaign worked brilliantly. By the end of the 1990s, America was producing and consuming more plastic than ever—most of it ending up in landfills. Even today, 60 percent of Americans believe that plastic is endlessly recyclable, when in fact less than 10 percent of the plastic you carefully sort and put in a blue bin is recycled even once. Whoever came up with the idea of branding plastic as "recyclable" was very clever. But I don't think they were smart. They took their talent—their understanding of the world and of people's motivations—and used it to trick well-intentioned people into thinking they were protecting the environment when in fact they weren't. That's their legacy.

Today, keeping consumers from learning the truth about plastic recycling has become, if anything, an even higher priority for oil and gas companies. As its profits are threatened in areas like transportation, the fossil fuel industry is counting on producing more plastic than ever to keep up. If you're a talented marketer or advertiser, you'll have plenty of chances to help mislead the public, and you'll probably be very well-compensated for doing so. But let's not mince words. If you're a messaging guru who figures out the best way to attack cleaner, cheaper energy while ignoring climate change, or a geologist who makes it cheaper to drill for offshore oil, or an investor who makes it easier for fossil fuel companies to afford to drill in remote areas, you're using your talent to make the world *worse*. Especially now, at a time when the consequences of those actions couldn't be clearer. There's no need to condemn anyone to weeping and gnashing of teeth, like in the parable from Matthew. But we also don't need to treat people who are squandering their talent as if they're doing something smart.

You know who I think are really smart? People who could easily use their talents for personal gain at the expense of doing good but instead use their gifts to see two moves ahead and make the world better.

By that definition, my friend John Podesta is one of the smartest people I know. He went to Georgetown Law, worked for the Department of Justice, spent years on Capitol Hill, served as deputy chief of staff and chief of staff for President Clinton, founded the most influential progressive think tank in the country, became a senior advisor for President Obama and then a senior advisor for President Biden. The issue he cares most about is climate—and he knows his stuff.

What's really striking about John is the way he combines his knowledge with strategy. Take the Paris Climate Accords. In President Obama's second term, getting a big international agreement on climate was a big priority. At the same time, since there was (as we learned the hard way) no guarantee he'd be replaced by a Democrat, the clock was ticking. It wasn't easy, but in 2015, after a last-minute boost from the president himself, representatives of countries around the world came together in a climate agreement that, while imperfect, was the most comprehensive and ambitious ever reached.

At least, that's the story most of us saw play out. But the full story goes back years. In the beginning of 2013, when John was named the White House's climate advisor, he began meeting one-on-one with ministers from all the countries who would need to sign a future accord. Knowing John, I bet he started building those relationships *before* he had an official White House job.

By the time the Paris Climate Conference was scheduled, America was ready to start putting the pieces together. Working closely with then–Secretary of State John Kerry, he figured out what China could live with. He figured out what India needed to get to yes. He knew what Mexico wanted most out of a final deal. When delegates arrived in Paris, they had an enormous head start. This isn't to take anything away from everyone else on the Obama team who helped get the accord across the finish line, including Secretary Kerry and the president himself, because that was still incredibly difficult. But it would have been impossible if the groundwork hadn't been laid, piece by piece, for years—and John Podesta was a huge part of that.

Compare John to another important person in the nation's capital, Senator Ted Cruz. Ted Cruz was valedictorian of his high school class. He went to Princeton and Harvard Law. Clerked for the Supreme Court, then argued cases before the justices and won. Got himself elected to the Senate. The guy has talent. But he also represents the state that already gets more of its energy from solar and wind than any other, and has more potential for the clean energy industry than any other, and he's completely in the tank for oil and gas. As recently as 2015, he argued that there had been "no significant warming whatsoever" since 1997 (the same cherry-picked 1998 data point, you'll notice, from the *Wall Street Journal* editorial page) and that carbon pollution was "good for plant life." He says stuff that just isn't true, things that would require him to either be ignorant of even the most basic facts or just not care that he's misleading his own Texas constituents.

By all accounts, Ted Cruz's brain has a lot of processing power. But what is he using all that brainpower for? A hundred years from now, the Paris Climate Accords will be remembered as one of the biggest steps we took on the journey to help keep our planet habitable. A hundred years from now, Ted Cruz will be remembered as another loud-mouthed senator who put his ego ahead of his country as the planet burned.

Of course, someone like Ted Cruz is an extreme example. Often, poor judgment is the result not of rampant cynicism or bad intentions but of narrow thinking. When people can't see beyond their circumscribed notions of the way things are, their ideas—and especially their predictions—become prone to major errors.

That's not just true in the fossil fuel industry today; it's been true for as long as I've been an investor. Back when I was in the mergers and acquisitions department of Morgan Stanley, a lot of our work involved takeovers in oil and gas. Basically, oil companies wanted to buy other oil companies, and we helped them figure out whether and how to do it.

I'm not especially proud of this work. It certainly didn't make the world better. But it did introduce me to oil price decks, which really matter if you want to understand energy and climate.

Oil price decks predict what a given barrel of oil will sell for in the future, so that companies and investors can compare that price against their projected costs and figure out how much money they'll make. The cost side of the equation is relatively straightforward to assess. The amount of oil in the ground can be measured, or at least estimated. Expenses for drilling vary by field, but they also can be assessed with some confidence. What fluctuates—a lot—is how much that barrel of oil will sell for on the open market, because most oil takes years to get out of the ground. So when oil companies buy each other, they're taking a huge commodity risk that they don't hedge out. They're telling their investors, including their shareholders, "In five years, we think the price of a barrel of this oil will be X dollars more than it costs to produce." That projection is an absolutely critical part of the deck.

As it happens, when I was looking through these decks, oil prices were really high and getting higher. From 1971 to 1981, they rose by 1,000 percent. You would think the conventional wisdom at the time would have been, "What goes up must come down," or at the very least, "This can't go on forever." But no. I remember reading one deck that said that oil would never stop going up. It projected that prices would go from about $30 per barrel in 1980 to $156 a barrel by 1983, and then up to $173 in the years after that. In the thirty years since then, even with inflation, the price of a barrel of oil has still never hit $150 a barrel. Not for a single day.

Whoever was behind that price deck must not have known a lot about economics, right? Wrong. I don't know who literally wrote the deck, but the firm it came from was run by Alan Greenspan, someone with such a sharp economic mind that he led the Federal Reserve for nearly twenty years straight. You may not agree with every one of his decisions, but

Alan Greenspan understands economics. He certainly knew that the price of a commodity can't rise forever, and it seems likely that anyone he hired would have known that, too.

So what happened? My guess is that he and his firm weren't able to break free from narrow, circumscribed thinking. Economic theory at the time was that oil prices were inelastic, or to put it differently, that demand for oil would be constant no matter how expensive it was. As it turned out, that was wrong. Oil prices are inelastic, but only to a point. Once the price gets high enough, the typical rules of economics kick back in: people use less, demand falls, and the price falls with it. Strangely, the fatal flaw in that price deck was not lack of expertise but lack of imagination. It couldn't conceive of a world in which circumstances changed in such a way that the prevailing theory ceased to be true.

We're still living in a world where people involved in the fossil fuel industry can't wrap their heads around the possibility of people using significantly less oil, even as the alternatives become more widely available and cheaper every year. But that's not the only part of climate where otherwise-smart people are hamstrung by narrow thinking. You'll remember that in our five-plus-one plan for stabilizing the planet, agriculture was one of the five big areas where we have to reduce emissions. But it's also an area where lots of experts engage in circumscribed thinking—and where a few incredible, smart, imaginative people are proving the old theories wrong.

One of the biggest myths about protecting the planet from climate change is that we can have a livable planet or we can eat food we like, but not both. The fossil fuel industry would love it if we had to choose between clean energy on one hand and hamburgers and bacon on the other. Because, let's face it, a lot of people would choose the burgers and bacon.

But it's a false choice. We don't need to sacrifice the food we like to eat. We need to change the way we *produce* those foods.

In the United States, we're really good at agricultural productivity. In 2020, the United States ranked third in agricultural output, behind China and India, even though we have far fewer farmers. But we're not good at growing healthy, delicious food in a sustainable way. Our most highly subsidized crops are wheat, corn, soy, and rice—and we turn most of these crops into processed food either for ourselves or for animals that we raise in awful factory conditions.

Not only is this kind of farming unhealthy, it's also harming the planet and putting human health at risk. To understand exactly why current methods of industrial farming are so harmful, you have to start with the land itself. When Europeans first arrived on the American continent, the soil here held a huge amount of carbon—according to some researchers' estimates, more than 10 percent by weight. One of the reasons for this was that most of the plants in the ground were perennials, meaning they grew back year after year. That gave them time to put roots deep into the soil, which helped fix nitrogen and prevent erosion.

Unfortunately, some people—especially those from the budding agriculture industry—looked at all this incredible farmland and treated it as yet another resource to extract. Their attitude toward our soil was basically, "This stuff is great! How do we use it up?" Agribusiness pressured farmers to replace perennials with annuals, which die after just one season. A shorter lifecycle means smaller roots, smaller roots mean more erosion, more erosion means more carbon-rich soil disappearing. On top of that, extractive farming tends to plant the same crops in the same places year after year, which drains nutrients from the soil that remains.

Basically, extractive Ag sees soil not as a complex, self-renewing ecosystem, but as mere fuel, and now that fuel is just about spent. The carbon-rich soil that Europeans found a few hundred years ago? It's gone.

Today, an average handful of American soil contains just 1 percent carbon. Throughout our country, we've replaced some of the best farmland on earth with some of the worst.

So how do we still produce so many calories? In large part, by spreading huge amounts of chemical fertilizer all over the place. And how do we make chemical fertilizers? Using oil.

Think about how nuts this is. We leached the carbon out of our soil, in a way that sent lots of it into our atmosphere. Then, to make up for the fact that our soil has been degraded, we manufacture a bunch of fossil fuel–based fertilizer, which sends even more carbon into our atmosphere, and dump it onto the ground. No wonder agriculture is responsible for so much greenhouse gas every year.

A lot of farmers, people who genuinely care about what they do and are experts at it, have been operating this way for so long—with enormous pressure from agribusiness and the encouragement of the US Department of Agriculture, which is constantly lobbied by the Big Ag companies—that they think they have no choice. The conventional wisdom says that the only way to feed the world's growing population is to stick with the status quo.

Gabe Brown used to think that way, too. In 1991, he took over his in-laws' 1,760-acre farm outside Bismarck, North Dakota. His wife's family had been practicing extractive farming—with all its chemicals and blunt-force approach to nature—since the 1950s, so he did, too. But in 1995 and 1996, massive hailstorms destroyed his crops. The next year, a blizzard killed most of his cows. Gabe was on the brink of losing his farm. Out of desperation, he decided to try something he'd heard about but had never put into practice: regenerative agriculture.

Regenerative Ag is about going back to the closed ecological loops that farming used to depend on, and doing it at a large enough scale that farmers can make a living and the rest of us can have enough food. In his book, *Dirt to Soil*, Gabe describes his journey. He cut out chemical pesticides. Then fungicides. He began growing multiple species together in ways that enriched the soil, and he allowed leftover biomass from each year's harvest to remain on the ground, where nutrients were reabsorbed by the earth for the following year. Before long, Gabe was able to cut back on synthetic fertilizers, and by 2010, he'd stopped using them altogether. He didn't need them.

But the most important change Gabe made wasn't in the way he farmed. It was in the way he thought about farming. As he describes it, he stopped thinking about soil as a medium for living things and started to think about it as a living thing itself. That's what Gabe means by "dirt to soil." And I think it's pretty darn smart.

Today, Gabe Brown's 5,000 acres—nearly three times more than when he almost went bust in 1997—are exploding with life. Meanwhile, right next to it is a conventional farm that, by comparison, looks like a field full of withered crops. In fact, Gabe has twelve times the farm revenues of his neighbors who still rely on fossil fuels.

By farming in a way that doesn't destroy the planet, Gabe is growing more food and making more money. He's doing exactly what the fossil fuel industry says is impossible to do. On top of that, he's putting carbon back into the ground. An acre of land on a conventional Northern Plains farm typically holds between ten and thirty tons of carbon in the top forty-eight inches of soil. An acre of land on Gabe's farm holds ninety-six tons—more than three times as much.

I admire Gabe. More than that, I've seen the way he's inspired Kat—who knows way more about agriculture than I do, as she runs what we call our "regenerative ranch." About ninety minutes from San Francisco, it's designed to show that sustainable agriculture can go beyond even crops, proving that you can raise and grow cattle—historically one of the worst animals to eat, from a climate perspective—in a way that isn't just carbon neutral but carbon *negative*. Not only that, but the meat you get from raising carbon-negative cattle tastes better.

I'm going to repeat that because you might not believe it. The fossil fuel industry tells you that you can't save the planet and have beef. In fact, if we grow beef the right way, it tastes better and *sequesters* carbon.

All across the country, farmers are breaking free from conventional wisdom to grow food in a way that tastes better, works better as a business model, and brings us closer to net zero in the process. In Eros, Louisiana, Donna Isaacs runs DeLaTerre Permaculture Farm. Guided by her farming upbringing in Jamaica, she started in 2019 with fourteen acres of depleted land, and in less than five years she more than tripled her farm's size to offer produce, meat, and eggs raised using regenerative practices. In Pescadero, California, Brisa Ranch—run by Verónica Mazariegos-Anastassiou, Cole Mazariegos-Anastassiou, and Cristóbal Cruz Hernández—grows a wide array of organic fruits, vegetables, and flowers while supporting their community and reducing dependence on fossil fuels for our food.

These examples don't in and of themselves prove that we can produce all the food needed to feed a growing population purely through regenerative means. But they clearly demonstrate that the range of possibilities is much wider than the extractive agriculture industry, or the fossil fuel industry, would like you to believe. We don't have to choose between feeding the planet and stabilizing the planet. By adopting more regenerative agriculture, and finding ways to scale its techniques, we can lower the carbon pollution associated with producing food while increasing the amount of carbon our soil can hold.

That's especially true when we combine the best practices of regenerative Ag with new technology. One example? Scientists have developed "green ammonia," which doesn't require fossil fuels to produce. If we can scale its production, we'll be able to cut nearly all the carbon pollution associated with manufacturing fertilizer.

As usual when it comes to clean energy, oil and gas is going to stand in the way of innovation. But they can't hold back the drive to take the best there is and make it better. Because that's what America does. We figured out how to fly across continents and put human beings on the moon and get humanity's accumulated knowledge into a device smaller than a pack of playing cards.

Ever since I've begun looking for them, I've met countless creative, smart people who are devoting their talents to something bigger than themselves. They're able to connect the dots and understand the world in brand new ways. And they're harnessing their breakthroughs to build movements, inviting others to join them in doing the impossible.

What's even more hopeful, as far as I'm concerned, is that we all have a chance to be smarter tomorrow than we are today. Everyone's got a different part to play in getting us to net zero and saving the world from climate catastrophe. We all have the opportunity to redefine smart together. This is a moment to ask ourselves life-changing questions. How can I help connect the dots? Where can I break free of narrow thinking and change the way people around me see the world? How can I draw not just on my own experience but transcend it to help others?

If you ask yourself these questions, I can't guarantee you'll become a US senator from Texas or one of our country's richest bankers. But you'll be worthy of admiration. You'll be the kind of role model we need more of today. Most important, you'll be the kind of person who is doing their part in the face of the greatest threat we've ever known.

And that's pretty smart.

CLIMATE PEOPLE

Shreya Dave, Brent Keller, Jeffrey Grossman

"Twelve percent of US energy consumption goes into separating different chemical compounds from one another in manufacturing raw materials. It's about the same as all the gasoline in all the cars and trucks in the US every year," explains Shreya Dave. "That kind of blew my mind."

As a PhD student working in the lab of MIT professor Jeffrey Grossman, Shreya's focus was far from chemical separations. Instead, it was on desalination, specifically ways to better separate sodium chloride from water to improve clean water access. Working with fellow student Brent Keller, Shreya helped design a new, more efficient water filter. Unfortunately, it was too expensive to be widely used. But after coming across an article in *Nature* about the amount of energy that goes into separating chemicals, and the potential to make this process better for the planet, they began experimenting to see if the technology in their filter could be applicable elsewhere. The result was Via Separations, a company started by Shreya, Brent, and Professor Grossman that's helping to decarbonize the manufacturing industry.

The methods for separating chemicals used in manufacturing are complex, but Shreya likes to think of them as similar to getting pasta from a pot of boiling water. Most companies' separations involve evaporation. That's like boiling off the entire pot of water to get to the pasta, which takes a tremendous amount of heat that typically comes from fossil fuels. Via, based in Massachusetts, has developed the equivalent of a pasta strainer for chemical separations—a process that can achieve the same results as evaporation while using 90 percent less energy.

Via's filtration system is vastly more efficient—which means that in addition to emitting less carbon pollution, it helps businesses bring down their energy costs. And because chemical separations are such an important part of modern manufacturing, the climate impact of Via's tech could be truly transformative: the Department of Energy estimates that switching from heat-based chemical separations to filtration-based ones could save roughly 3 billion tons of carbon every year. That's the equivalent of taking *every single one* of the world's passenger cars off the road.

"A lot of what you hear about tech in the climate world is very trendy and very flashy, and what we're doing is really kind of grungy," says Shreya. But for her, that's just fine. After four successful pilot demonstrations, Via is starting to work with businesses to install its filtration system on a wider scale. While new ways to separate chemicals might not get the headlines other breakthroughs do, this hugely important transition is closer to becoming a reality every day.

CHAPTER VII

AGAINST FOOTPRINT SHAMING

According to the *Chicago Tribune,* Ruth Eckert was really good at growing vegetables. When World War II broke out, her eleven-year-old son Allen, who had spent a summer on a farm, wanted to be part of the war effort. So, despite being, as the paper described it, "in ill health," she accepted an offer from a Melrose Park family of some land for a victory garden—even though the nearest available plot was thirteen miles away.

Through the spring and summer of 1942, Ruth and her two sons took a streetcar for eight miles, then walked another five miles to reach their tiny patch of land in Melrose Park. They'd tend to their garden, then trek the five miles back to the streetcar station before taking the eight-mile ride home to the North Side of Chicago. In the fall, they harvested, hauling baskets full of tomatoes, corn, cucumbers, and rutabagas back home. In a glowing article, the *Tribune* reported that the Eckerts canned at least 200 quarts of vegetables.

The Eckerts weren't alone. You've probably heard of these victory gardens, which sprung up across the country during World War II, built on a similar nationwide effort from World War I. The story goes that to protect the United States from shortages, and to save meat and other high-protein foods for the soldiers on the front line, millions of Americans dug, planted, and harvested. Victory gardens popped up in backyards, in vacant lots, and on rooftops—Eleanor Roosevelt even planted one on the White House lawn. Sales of pressure cookers soared as canning became ever more popular. In 1943, these patriotic home gardeners produced about 40 percent of the fresh vegetables consumed in America.

But did victory gardens lead us to victory? The short answer is no.

In a *New York Times Magazine* article from 2020, Jennifer Steinhauer writes that despite the imposition of rationing, there was never any real risk of food shortages in America. As for saving meat for the troops, she notes, "In 1942, Americans consumed 138 pounds of meat per capita, a mere three pounds less than the prior year." Allan M. Winkler, a history professor at Miami University of Ohio, was more explicit: "Victory gardens were a symbol of abundance and doing it yourself, but that was more symbolism than reality."

So, was Ruth Eckert a hero for canning 200 quarts of vegetables in a year? Or was she just a well-meaning Chicagoan with a green thumb who, despite her hard work and good intentions, had little real impact on the war effort? When it comes to defeating a global threat, how much does individual behavior really matter?

That last question is the important one for climate people. Because while the global threat we're facing is different from the one America overcame in World War II, we still need to figure out what role individual behavior needs to play. Today, the way we tend to measure one's individual effect on climate is not by the number of rutabagas harvested, or pounds of canned goods pressurized into service. It's by the size of your carbon footprint. To win on climate, we need to think about our individual carbon footprints in a new—and much more useful—way.

You might reasonably wonder if I'm saying this to justify not caring about my own carbon footprint. It's actually the opposite. For me, as it does for many people, being conscious of my own personal carbon emissions begins with consumer choices. I have solar panels on my roof. I drive an electric car. I also make a choice that, while hardly heroic, sets me apart from most people in my circumstances: I only fly commercial. There are two reasons for this. First, private planes are incredibly elitist. To fly on one is to declare yourself independent of the kinds of problems everyone else—even fairly well-off people—have to deal with. That makes me uncomfortable. But the more important reason is climate. Private jets aren't just more wasteful than commercial passenger planes, they're *wildly* more wasteful. On average, they release twenty times more greenhouse gas per passenger. I don't think that my being inconvenienced is worth doing that much additional damage to the world.

The biggest factor in my own carbon footprint, however, is probably the regenerative ranch. Because of the agriculture techniques employed there, the ranch absorbs tremendous amounts of carbon that otherwise would remain in the atmosphere. I fully recognize that most people can't start what amounts to their own personal carbon sequestration program, but in my case, if you add everything together, I'm pretty sure that I don't just have a carbon-neutral footprint. I have a carbon-negative one.

So now, having established my victory-garden bona fides, let me explain why obsessing over your carbon footprint isn't productive. Actually, it's worse than unproductive. It's actively harming the planet.

Before we get into the details of carbon footprints, it's worth looking at the history of the term. The idea of an ecological "footprint" has been around since the 1990s. But the term "carbon footprint" only entered widespread use in the mid-2000s, after it was the centerpiece of a $100 million advertising campaign—for an oil company. "What on earth is a carbon footprint?" asked British Petroleum, as part of a push to rebrand itself as environmentally friendly. The company also unveiled the first widely used carbon footprint calculator, inviting the public to, in the words of journalist Mark Kaufman, "assess how their normal daily life—going to work, buying food, and (gasp) traveling—is largely responsible for heating the globe."

BP hasn't gone "Beyond Petroleum," as their advertisements have it. Twenty years after their ad blitz, they still make obscene amounts of money by pumping oil out of the ground. So why would a fossil fuel company be so eager for the rest of us to start calculating our footprints?

One answer is that focusing on the fact that most people have a carbon footprint, even if it's a small one, makes it easy to impose a kind of purity test: unless you make perfect choices, you don't get to be taken seriously about climate issues. The fossil fuel industry and their media allies love purity tests, because having an all-or-nothing standard allows them to dismiss climate people without having to engage with, or listen to, the arguments they're making.

One of the best (or worst) examples of this involved John Kerry. I first met John when I helped raise money for his presidential campaign in 2004. I admire him and think he would have made a great president. Of course, things didn't work out that way. But thankfully, John didn't fade from public life. As Barack Obama's secretary of state, he's probably the person, along with John Podesta, most responsible for getting the Paris Climate Accords across the finish line (other than President Obama himself). After Joe Biden won the presidency, John joined his administration as climate envoy. Like Al Gore, he's a quintessential example of an important national figure who recognized the dangers that come with burning limitless quantities of fossil fuels and threw himself into the fight.

So imagine the oil and gas industry's delight upon learning that in 2019, when John flew to Iceland to accept a prize and deliver a speech on climate, he did so on a private jet. *National Review* accused him of "insufferable hypocrisy." Fox News had a field day. Newt Gingrich tweeted,

"Do you really expect a political Prince like Kerry to sit in an Icelandair seat with mere citizens?"

As John pointed out at the time, these attacks were obviously made in bad faith. He buys carbon offsets; his climate work means his schedule was much busier than the typical person's; showing up in person offers a chance to build relationships that can't be replicated remotely. Of course, none of that mattered to his critics. What actually made the right-wing media so gleeful about John's flight was the opportunity to claim that because he flew private instead of commercial, everything else he's ever done or will do on climate doesn't count. And that claim is just silly.

Let's dive into it. A private jet from Boston to Iceland and back again releases about 21 metric tons of carbon into the air. As John pointed out, he'd purchased offsets, so in the final accounting, the emissions that resulted from his trip were substantially less than that. But just for the sake of argument, let's say he hadn't done any offsets, or that the offsets were ineffective. The Paris Climate Accords, which John played a crucial role in negotiating, are expected to cut emissions by 500 million tons annually. That's the equivalent of more than 22 *million* round-trip private jet flights to Iceland per year. And that doesn't even count all the work John's done that had nothing to do with the Paris agreement, including as President Biden's climate advisor. He has been an incredible climate advocate.

Nobody has a perfect carbon footprint. But that doesn't change the fact that John's devoted decades of his life to trying to protect our planet out of a sense of public service, while right-wing media organizations team up with the fossil fuel industry to destroy our planet for their own personal profit. I just want to say to these guys, "Give me a break! John's a hero." Adding to your carbon footprint isn't something to brag about, but bad-faith footprint shaming is deliberately destructive. And footprint shaming is widespread, not just by the fossil fuel industry. For years, I argued with the administration at Stanford about whether they should divest from oil and gas. My side of the argument was simple: Stanford lists sustainability as one of its core values, and if they truly value sustainability, and want the energy transition to happen as fast as possible, then investing in fossil fuels is a direct contradiction of that. Also, Stanford is one of the top schools in the world, with one of the world's largest endowments. If Stanford stopped investing in fossil fuels, it would send a signal to other universities and institutional funders that you don't need to root for the end of the world to generate returns.

I love Stanford. Many of the smartest, most inspiring people I know—including in the climate world—do their work and research there. So far, however, even though my alma mater is doing an enormous amount of good work on sustainability, they haven't divested from fossil fuel companies. They believe that fossil fuels meet real-world, essential needs (which is true), and, therefore, are necessary parts of today's economy (also true). But they invest in companies that are intent on preserving and profiting from this destructive status quo and trying to keep us addicted to fossil fuels for far longer than is necessary, healthy, or safe.

At one point, I got fed up and pointed this out to a senior administrator at the school. At which point, this professor with a PhD, helping to lead what might well be the premier university in the wealthiest country on earth, turned to me and said: "I'll start listening to these kids on climate when they stop driving their cars."

As this person fully understood, in most of the US, not driving a car means not being part of the American system. Yet he was saying that if you don't have a perfect carbon footprint, for reasons largely beyond your control, you don't get to have a voice about saving the planet. It's a very convenient standard to set if you're looking to avoid examining a question. But it's also completely unfair.

Footprint shaming has become so ingrained in our culture that I've even heard climate people shame themselves. Not too long ago, I was on a call with one of the board members of the Rocky Mountain Institute, a nonprofit that helps businesses take action on climate without sacrificing their competitive advantages in the marketplace. RMI does exceptional work. It's one of the NGOs I admire most. Their former board chair, Ted White, is a great friend of mine, and I was proud to join them to create the Global Cooling Prize, which helped jump-start more efficient HVAC and air conditioning systems. But on that phone call, the board member, whom I'll call Dan, sounded despondent.

"We're never going to solve this," he said.

"Okay," I replied, a little taken aback. "Why do you say that?"

Dan explained that he had just flown to New Zealand and back on a week-long vacation. "I'm on the board of the Rocky Mountain Institute, and I still fly across the world and burn all these fossil fuels. Are you really telling me that the other 8 billion people in the world are going to be more responsible? I don't think so."

Whether Dan should or shouldn't have taken the trip to New Zealand is up to him. But to feel that the planet is doomed or that he's unworthy of being part of the solution just because he flew on an airplane would be like telling people during World War II that they were not really anti-Nazi unless their victory garden was the size of Ruth Eckert's. Dan's a dedicated, thoughtful guy helping our society solve one of its toughest problems. If anyone should feel good about his life's work, it's him.

Turning a collective problem into a matter of individual responsibility is also exactly what the fossil fuel industry is trying to do. If every climate activist who ever made a non–carbon neutral decision lost all credibility, or became wracked with guilt, the oil and gas companies would win easily. We need systemic change, not perfect people. Because perfect people don't exist.

The good news is that because of how fast the technology is changing, the systemic change we need has never been closer to reality. The climate-person sales pitch isn't, "Give up the chance to go on a trip with your family." It's, "Let's bring down the cost of green hydrogen to where it's *below* the cost of jet fuel. Even if you don't care at all about protecting the planet, who wouldn't want to spend less on plane tickets?"

Ultimately, my biggest concern with climate guilt is that it's a distraction from the work we need to do. I'm all for making greener personal choices, but we don't have time to waste feeling ashamed about the choices we made that could have been a tiny bit better. Being on the right side of history—changing your life to help humanity deal with the greatest threat it's ever faced—is hard enough without self-recrimination. Don't let the oil and gas companies trick you into feeling footprint guilt.

Besides, individual footprints are just one small piece of the puzzle. We need to stop obsessing over the question, "What's the size of my carbon footprint?" and focus on a much bigger question:

"What kind of impact will my life have on our planet?"

Instead of our goal being to have carbon-neutral footprints, it should be to lead carbon-negative lives. The best way—perhaps the only way—to do that is through collective action.

For several years, I've been involved with one of the more direct forms of collective action available to Americans: ballot measures. In twenty-six states, organizations and individuals can sidestep the worst parts of partisan politics and put specific-issue questions directly in front of voters. Californians are especially used to sorting through dozens of propositions each election year, but many states allow for these kinds of issue-based campaigns.

Successful ballot measure campaigns are about doing big things together. First, people have to come together to get the issue on the ballot, which requires lots of signatures from all over the state. Then they have to persuade their fellow citizens that the measure is worth passing. If, in the end, the measure does pass, it represents a shared commitment—a mandatory change to the entire system, not just a suggested change in individual behavior.

Ballot measures are especially useful for climate, because outside of Washington, DC, when you get past politicians and their wealthiest donors, climate is not nearly as political as you might think. It's not just Democrats who care about it; poll after poll shows that Republicans do, too. Unsurprisingly, the more the natural world changes, the more the percentage of Republican voters who would like to see our government do more to protect us from climate change grows.

The biggest difference between Democratic and Republican voters is not whether they care, but how much. A lot of Democrats think climate is one of the most important issues facing the country today. It's usually ranked in the top three, and almost always in the top five. Republicans, on the other hand, almost never rank climate as a particularly high priority. This, of course, will also change as the planet changes.

Republican politicians who prop up and bail out the oil and gas industries clearly aren't doing what their constituents want—but as long as they do all the other things their voters want more, on issues like taxes, guns, and immigration, they're unlikely to pay a political price. This doesn't mean that passing new laws is impossible—far from it—but it's more difficult.

Putting climate directly on the ballot, then, lets people voice their opinion on an issue rather than on a political party. Which is why, after helping to lead a winning campaign for a clean-energy California ballot measure in 2010, I started to think ballot measures could work in other places, too.

In 2018, we put forward ballot measures in Nevada, Michigan, and Arizona. The proposition was a very simple yes or no: should the state generate 50 percent of its power from clean energy by 2030? We could have restricted our campaigns to blue states, which probably would have increased our odds of winning. But we deliberately chose three purple states, because we wanted to prove that climate was a winner throughout the country, not just in places where Democrats already did well.

We ended up winning in all three states. But it's worth highlighting that we won each campaign in a different way, through a different model of collective action that drives system-wide change.

In Michigan, we met with the two big utilities to explain why "50by30" wouldn't hurt them, and in fact would grow their businesses. Then we went to the United Auto Workers, one of the most organized groups of workers anywhere, and one of the most influential political forces in the state. They were forward-thinking enough to realize that to lead the world in making electric vehicles, Detroit couldn't get all of its own electricity from coal. They said to us, "We'll support it." They sat down with the biggest utilities in the state and agreed that instead of waiting for a ballot measure to force them to change, they'd go ahead and commit to our 50by30 goal.

Our campaign in Nevada was similarly uneventful. The big utility there was owned by Warren Buffet's company, Berkshire Hathaway, and while they refused to publicly support the measure, they didn't oppose it either. With no coordinated opposition, we won by a wide margin.

The third state, Arizona, was a different story. The utility there was run by someone who hated solar energy. He told us, "We have six percent solar, and that's the most we can possibly do." From the standpoint of economics, not to mention common sense, that was obviously false. Arizona has almost no oil. It has hardly any gas. You know what Arizona has a ton of? Sun! Basically, this guy was in favor of buying dirty energy instead of getting unlimited, free, clean energy.

That happens all the time on climate. People get sucked so deep into the oil-and-gas bubble that they'll fight against their own community's interests. We tried to show the utility all the reasons that expanding solar would create jobs and help their business, but this person had no interest in listening. Instead, his utility went to the Arizona attorney general and gave him more than $400,000 in campaign donations, a huge sum of money for someone running for attorney general in Arizona. Then, at the last minute, the attorney general changed the ballot language to say that the state would transition to renewable energy "irrespective of cost to consumers." The new language strongly implied that if the ballot measure passed, everyone's electricity bills would go up. Of course, that wasn't true, and in many cases consumers' costs would go down. But it didn't matter. With the new language, we were doomed.

But we didn't give up. After we lost in 2018, we went back to the utility and said, "You may have won this round, but you haven't won the fight. Two years from now we'll be back." As it happened, we never got our rematch with the utility, because they got caught up in an FBI corruption investigation. Knowing that we were going to put our question back on the ballot—and that it would be much harder to get a last-minute change to the language this time—the utility decided to commit to 50by30 on their own.

Each of these models—working with an established organization, defusing potential opposition in advance, and beating the opposition through persistence and the faith that the truth will ultimately prevail—is an example of how collective action can work, and how it can lead to large-scale change. Today, all three states are well on their way toward having 50 percent clean energy by 2030. Nor will it surprise you to learn that despite the dire warnings we received in Arizona, the utility's been able to handle far more than 6 percent solar energy.

Just as important, we proved that we don't just have to rely on individual behavior to usher in the post–fossil fuel era. We can do it together, by changing the entire system that dictates how energy is created and deployed. In fact, that's exactly what Americans—even those in states bitterly divided by party—want us to do.

With all this in mind, let's return to the question of how climate people should treat individual carbon footprints. But instead of obsessing over personal consumer choices that are often the result of economic or societal pressures and ultimately don't make a significant impact one way or the other on climate, let's use the lens of collective action. How do one's individual choices help increase the chance not just that other people change their personal behavior, but that we change the systems that affect *all* behavior?

This is a much better question than whether someone should feel guilty for driving a car. When you ask it, you start to see that there are some good reasons for growing a victory garden after all—as long as you turn your personal actions into a source of inspiration rather than one of needless shame and guilt.

First, thinking about how much carbon you personally are responsible for is a commitment device. It ensures that devoting part of your time on this earth to stabilizing the planet—in other words, to being a climate person—is always at least partly on your mind. No less important, your victory garden can be a public statement of values and make others feel less alone. Helping someone else feel a sense of community and letting them know that they're part of something bigger than anything the oil-and-gas industry can match can be the difference between that person finding the strength to double down on their efforts or give up. We're up against some of the most influential and best-funded media, financial, and political interests in history. It can feel like a lonely fight. Through victory gardens, we can remind one other that we're in this together.

Second, making small changes in personal behavior—changes that in and of themselves don't do much toward getting us toward net zero—can spark something much bigger. In 2011, Buzz Smith, a business manager at Apple, was on his way to work when his car was rear-ended and totaled. Most people would have filed an insurance claim and bought a near-identical replacement, but Buzz and his wife were already thinking about climate, even powering their home by wind energy. Not long after the crash, they switched to a fully electric vehicle, and they soon realized their decision hadn't just helped the planet—it was saving them $185 a month on gas.

Buzz became, in his words, an "EV-angelist." Today, he's the director of dealer outreach at the Texas Electric Transportation Resources Alliance, helping to sell electric vehicles in the heart of the American fossil fuel industry. When he gives talks to car dealers (often wearing a sparkling red tuxedo, for effect), his big piece of advice for getting EVs off the lot is simple: "Do not mention the environment. Do *not* mention climate change." Instead, he lets potential customers know how great the products are—how much they'll save on gas and maintenance, how much they'll get back in tax credits, and how the current generation of EVs outdrives its gas-powered counterparts. In other words, what started out as Buzz trying to reduce his personal carbon footprint turned into a mission to drive down pollution at scale while showing people that greener products can be cheaper, and better, than their alternatives. And if he can do that in Texas, it can be done anywhere.

Which brings me to the final, and perhaps most important, way a climate victory garden can help in the larger effort to stabilize the planet: the power of example. The value of setting an example is hard to measure, but whether you're talking about leading in business, government, or anywhere else, it can shape—or destroy—a culture. In other words, one's actions can ripple out until individual behavior isn't so individual after all.

I'll never be as good at growing vegetables as Ruth Eckert. I'll also never make perfect choices on climate—just as I'll never be a perfect person. So from one imperfect person to another, let's grow our victory gardens but stop obsessing over them. Let's live out our values day to day while always keeping an eye on the bigger picture. Let's recognize that we can set an example for others, not spend time on needless guilt when we fall short, and have the ability to spur our communities, our states, and even our countries to take the collective action that can change the world.

CLIMATE PEOPLE

Marika Ziesack and Shannon Nangle

Today, electric cars ride as smoothly and accelerate as quickly as cars that run on fossil fuels, while saving drivers money they'd otherwise spend on gas. Solar panels power our homes at a lower cost than coal or natural gas. But while agriculture is a huge driver of climate change, it's much tougher to find alternatives for many of the most delicious things we eat.

"If you've ever had vegan cheese, you probably know how awful it is," says Marika Ziesack, a co-founder of Circe, a manufacturing company working to develop climate-friendly things like plastics, biofuels, and foods. Some people may disagree with her, of course, but if you look at the size of the vegan-cheese market, you can see that many people share her opinion. In Marika's telling, vegan cheese is bad in part because it's missing the milk-based fats that help give cheese its flavor and feel.

If they were only focused on getting individuals to reduce their carbon footprints, Marika and her co-founder at Boston-based Circe, Shannon Nangle, might spend their time telling people to eat vegan cheese for the sake of the planet, regardless of how it tastes. But they know that's not how human nature works, especially when it comes to food. Instead, they have a simple thesis: if climate-friendly foods had the fats we know and love, people would eat more of them.

Marika and Shannon are unlikely partners. Marika grew up in Germany and at one point considered a career as a stuntwoman. Shannon is from Florida and thinks living on Mars might be a key to the future of humanity. They met in a lab at Harvard working on synthetic biology, the making or tweaking of things found in nature. It's this science that is powering their company.

Circe isn't the first company trying to make foods that are better for the planet taste better. But what I think makes it stand out—and the reason we've invested in it—is that it's tackling the problem in a creative and ambitious way. Marika and Shannon have developed microbes that eat carbon dioxide, sort of like how plants absorb CO_2. When combined with hydrogen and oxygen, a fermentation process allows these microbes to convert the carbon dioxide into triglycerides, the building blocks of the fats that give foods their distinctive properties.

Consider chocolate. Because they need lots of sunlight, rain, and humidity, cocoa beans—which are currently the only way to produce cocoa butter at scale—can only be grown in certain tropical climates, often leading to the deforestation of large tracts of jungle. This, combined with the fact that harvested beans and processed butter need to be shipped long distances, makes chocolate among the plant-based foods that result in the highest amount of carbon pollution.

Circe aims to make delicious, climate-conscious chocolate possible. In fact, because their microbes can take CO_2 from the air and turn it into cocoa butter, they may soon be able to make chocolate with a manufacturing process that sequesters more carbon than it releases. In other words, you might just be required—as part of your responsibility toward our planet—to add more chocolate into your diet.

In addition to climate-friendly chocolate, Circe is working to recreate the fats in foods like palm oil, beef, and, yes, cheese. With beef and dairy products among the highest contributors to agriculture emissions, an industry that accounts for more than 10 percent of global carbon dioxide pollution, Circe's CO_2-based microbes have the potential to lower greenhouse gas emissions by hundreds of millions of tons a year.

CHAPTER VIII

TAKE THE RIGHT-SIZED SWING

n 1975, I met Fleur Fairman. Her brother Frank played on the soccer team with me, and although she was a class above me, we ended up being friends with some of the same people. Before long, we became friends, and a few years later, after I started Farallon, she joined my great friend Katie Hall and me as a partner.

To say Fleur was a good investor was an understatement. She could anticipate events like no one else. Definitely better than I could. Lots of people claim they can put their emotions aside in a crisis and perform a careful analysis, but Fleur was one of the handful who actually could. She consistently took such smart positions that she didn't just beat the market, she beat the rest of our fund, often by a mile. In 1988, for example, the market was bouncing back from a historic crash, and Farallon did really well. Our return that year was something like 35 percent. Fleur's return was 100 percent.

But what really struck me about Fleur was that if she was confident an investment was going to do well, she didn't just act—she acted in direct proportion to her confidence. In fact, this was such an important part of her approach to investing that it drove her crazy when other people (me, for instance) failed to adopt it. On one occasion, for example, some people within the fund wanted to make a dicey bet. After some debate, we decided to pass. As it turns out, we made the right call. When the investment went bad, I proudly shared the news with Fleur.

"That thing blew up yesterday, and we own none of it," I said smugly. If I'd thought she'd join in the celebration, I was sorely mistaken.

"You could have made a ton of money, Tom," Fleur said. "If you were so smart, why didn't you short it?"

On other occasions, when I'd announce to Fleur that an investment had come good, as we thought it would, her reaction would be just as harsh. "You took a tiny position in something that worked out well. In other words, you missed taking a big position."

Fleur could be brutal in her honesty, but she was 100 percent right, and she was making a point that goes way beyond investing. You've probably heard the idea that quality is better than quantity. But that's not always true. Whether you're trying to seize opportunities, solve a difficult problem, or protect yourself from potential dangers, it's not enough to have a high-quality strategy. Quality needs quantity. You need to take the right-sized swing.

So often in life, well-meaning people ask themselves a question about direction: "Am I doing the right thing?" But as Fleur understood, we can't end there. We have to ask ourselves another question that is just as important, or sometimes even more so:

Are we doing enough?

As a movement, and as individuals, climate people need to ask that question more often. And if we want to protect our planet, and preserve human life as we know it, we need to be brutally honest with ourselves about the answers. There's a saying, often attributed to the legendary humorist Will Rogers, that makes a point not dissimilar to Fleur's: "Even if you're on the right track, you'll get run over if you just sit there." Even among people who know the science, worry about the dangers that come with a warming planet, and agree that action must be taken, I sometimes feel that there's too much applauding ourselves for being on the right track, while failing to notice that we're basically just sitting and waiting to get run over.

One of the most obvious ways in which we could do more involves money. I'll never forget something that happened at the 2021 UN Climate Change Conference in Glasgow. At the time, Raj Shah, who has served as the president of the Rockefeller Foundation since 2017, was trying to help solve one of the thorniest geopolitical problems related to climate change: how to get electricity to people who don't have it, without destabilizing the planet even further.

More than 700 million people around the world still have no access to electricity. To get to net zero by 2050, we don't just need countries to transition to electrical energy, we need to get those 700 million people on the grid using clean energy. That's a much less expensive proposition than it used to be—the cost has fallen even since 2021, and we're clearly at a point where it's cheaper to expand access to electricity via renewables than fossil fuels. But not everyone recognizes that, and in some cases, the quickest option for poorer countries would be to use fossil fuels to expand access to electricity. It's entirely understandable that countries would want to make the conventional choice—especially since they're responsible for so little of the carbon pollution humans have emitted—but the unavoidable fact is that it would be a disaster for our planet.

Raj is one of the most impressive people I've met while working in climate. After growing up in Detroit, he got a business degree from Wharton and an MD from the Perelman School of Medicine at Penn. Before moving to the Rockefeller Foundation, he served as Chief Scientist at the Department of Agriculture, and then became the head of the US Agency for International Development under President Obama. Using his considerable talent, influence, and connections, he set to work securing funding commitments from wealthy countries and other big foundations, pledging to help bring electricity to those 700 million people using renewable energy instead of oil, gas, or coal. By the end of the conference, he'd gotten $10 billion to create the Global Energy Alliance for People and Planet, a partnership supporting clean energy development in Africa, Asia, Latin America, and the Caribbean.

The partnership got a lot of press, and for good reason. It's incredibly impressive that Raj was able to raise so much money so quickly, and the funding he was able to secure will have a direct impact on nations and on people's lives. My concern, however, was that some people seemed to be celebrating as though we'd reached the finish line rather than taken an important first step. I felt a little bit like Fleur must have when I told her we had taken a tiny position in a good investment. I remember thinking, *Do you have any idea how much it will cost to bring electricity to 700 million people? Ten billion dollars is a big deal, but we have an opportunity to do so much more.*

In the years since 2021, more donations have flowed toward climate causes—but change is still not coming nearly fast enough. As Al Gore pointed out recently at the World Economic Forum in Davos, just 2 percent of global philanthropy goes toward nature and climate. A threat endangering all of humanity—and one we know how to solve, if we only had enough resources—gets just two cents out of every dollar donated to charity.

Why aren't we doing more? One explanation involves a concept I thought about frequently as an investor, which I like to call "the range of reason."

The easiest way to illustrate the range of reason is to describe the way I cross the street versus the way my kids do. I grew up in New York City, and while traffic lights exist there, New Yorkers have their own de facto method for determining if it's safe to cross: they look around. If they're going to get hit by a car, they don't cross. If they're not going to get hit by a car, they do cross. New Yorkers have things to do and no time to waste.

The way I cross the street drives my four children crazy. They were all born and raised in California, where people have the utmost respect for crosswalks and lights. Even at moments when the streets are empty, moments when it is totally obvious, to anyone, that no cars are coming, my kids will wait for the walk sign to cross, and when I don't wait, they get mad.

"Dad!" they say, "it's a red light! We're not supposed to cross."

The range of reason is that gap between what society expects you to do and what makes sense given the circumstances; between what conventional wisdom says is reasonable and what actually is. Don't get me wrong, I don't believe you should just write your own rules all the time (and I'll discuss rules in much more detail later). But I do believe your actions should be dictated by circumstances, not convention. You need to trust your judgment and have the courage of your convictions not only when it comes to what to do, but how much to do.

If you're afraid of seeming unreasonable or straying too far from the herd, you'll inevitably hedge your bets. That's exactly what has happened for decades when it comes to climate projections. For the past thirty-five years, the most frequently cited authority for predicting the trajectory and extent of global warming has been the United Nations' Intergovernmental Panel on Climate Change, or IPCC. Their assessments synthesize data from thousands, or even tens of thousands, of peer-reviewed studies. In terms of direction, they've gotten it right. IPCC reports have consistently predicted higher average temperatures, and average temperatures have indeed risen over time.

When it comes to magnitude, however, the IPCC has consistently missed the mark—and always in the same direction. In 1990, the IPCC estimated that global temperatures would rise between 0.1 and 0.3 degrees Celsius per decade—by 2001, the temperatures had already risen

0.5 degrees, and the decade-over-decade increases have only grown from there. That same year, the panel predicted between 0.4 and 1.1 degrees of warming by the end of 2025. We've already experienced 1.2 degrees of warming, with more than a year to go. In 2014, the IPCC's report said that the worst-case scenario for global temperatures was a 1.8 degree rise by 2030; in 2023, they put out a new report that said that 1.8 degrees of warming was now the most likely scenario, not the upper bound.

While it's hard to make accurate projections, if the error were random, you'd expect the IPCC's estimates to be too low about as often as they're too high. But that's not what's happened. Instead, over and over again, the rate of climate change either exceeds the IPCC's assessments or comes in at the upper end of their range. They've gotten lots of things right. But when it comes to the topline numbers—the ones that get the most attention—there's a bias toward understatement. Every time I finish reading one of these reports, it always seems toned down to me.

Meanwhile, other researchers and scientists have been significantly more accurate. In 1989, a report from the Worldwatch Institute think tank predicted that global temperature would rise between 1.5 and 4.5 degrees Celsius by 2050. Back then, the *Los Angeles Times* called the prediction "apocalyptic," but it now seems extremely likely to prove accurate. Johan Rockström, one of my climate Mr. Darcys, has used his position at the Potsdam Institute to say, for years, that climate change was far worse than most people realized. Like most predictions, these were far from perfect—among other things, researchers underestimated the impacts of extreme weather that have already resulted from climate change. But when it came to predicting the rate at which the planet was warming, they got it mostly right.

How is it that plenty of researchers consistently predict warming correctly, while what is quite possibly the world's most influential group of climate prognosticators keeps underestimating the speed at which our world is changing?

I think the answer is pretty simple: this is what happens when physics is done by politicians. The climate scientists who compile data for the IPCC may be brilliant, but they're not in charge of the UN. The UN is run by people who work for governments, and to be taken seriously as a government minister, you have to seem rational. So you tend to hedge. You try to alert people to the seriousness of the problem, but you don't want to sound alarmist. You hold back. You stay within the range of reason.

My suspicion is that, whether or not it's explicit, there's pressure on the IPCC to be just a little more optimistic than the data warrant. I understand the instinct to downplay a threat. You don't want to freak everyone out. But we owe it to ourselves, and to future generations, to be honest about the scale of the danger we face—even if it makes some people uncomfortable. Successfully addressing climate change is going to require us to step outside the range of reason when the circumstances warrant it, to stop viewing the world through rose-colored glasses just because we think that's what we're "supposed" to do.

There's another idea that relates to the range of reason, and that is to assume the possibility of chaos. In a strange way, just like my attitude toward jaywalking, my understanding of chaos has a lot to do with growing up in 1970s New York. I never got into much trouble when I was a kid, but I played a lot of sports in the park. New York back then was a rough city—much rougher than it is today—and people got in fights all the time during pick-up games. One day, on the basketball court, a big guy playing really aggressively was picking on a smaller player. As the game went on, the bigger guy got more and more obnoxious. Suddenly, the smaller guy pulled out a knife, put it to the bigger guy's throat, and said, "Cut it out."

The guy backed off. The other guy put his knife away. Eventually both guys left. But looking back, what's so strange to me is what happened next: we kept on playing. It wasn't that any of us had *expected* someone to pull a knife on another person over a pick-up basketball game. It was a totally insane thing to do. But if you were a teenager in New York in the 1970s, you took it for granted that totally insane stuff happened all the time.

For decades, people have been dismissing climate predictions as alarmist or unreasonable because they can't wrap their heads around the global equivalent of a guy pulling out a knife at a pick-up game: Greenland melting. Palm trees in the Arctic. Ninety-nine percent of coral reefs vanishing. A million species of animals going extinct. Coastal cities like Miami and New Orleans disappearing by the turn of the next century as sea levels rise. I'm not saying all these things will happen. I'm saying that just because they're hard to imagine doesn't mean they won't.

In fact, the best available science indicates that these are distinct possibilities. Greenland's ice is already melting seven times as fast as it was in the 1990s. Sediment samplings have found that during previous periods of heightened global temperatures, palm trees really did grow in the Arctic Circle. There's no reason to despair; not yet, anyway. With each passing year, it becomes even clearer that if we act at a scale that matches the size of the problem, we can avoid these tragedies. But we won't do enough if we refuse to accept just how bad things might get. When it comes to climate, assuming the possibility of chaos means being honest about the full range of what might happen if we don't act quickly enough.

In my experience, stepping outside the range of reason and assuming the possibility of chaos can be a pretty lonely experience. As an investor, it can be unsettling to hold a position—even a position you're confident in—when you know that everyone else thinks you're wrong.

As I learned the hard way, that's even more true in politics than in investing. For most of Trump's presidency, I watched people in the private sector whom I'd known and admired for years avoid standing up to him because they didn't think it was in their interest. Some praised him. Others tried to ignore him, or found ways to take both sides. It didn't take me long into the Trump presidency to reach a very different conclusion: that this man was un-American, that his actions went against everything the Constitution stood for, and that if he got re-elected, he'd try to destroy our country. I knew lots of people who agreed about the threat President Trump posed, but I didn't see an organized and continuing effort that reflected the extent of the danger our country was in. In October 2017, I did what is probably the most unconventional thing I've ever done: launched an organization called Need to Impeach.

Outside Washington, we gained traction almost immediately. We kicked off our campaign by releasing an ad highlighting some of the blatantly unconstitutional things Trump had done or threatened to do, like saying he would shut down news organizations that reported on him. Ten days later, our petition calling for impeachment had more than a million signatures. Over the next year, we held town halls and rallies across the country pointing out the ways in which Trump was a threat to democracy, and we ran ads making sure that voters in the 2018 midterms understood what was at stake.

Within the political world, though, people absolutely hated Need to Impeach. One of my good friends, a fellow donor to Democrats, told me I was being delusional. David Axelrod, President Obama's messaging guru, called Need to Impeach a "vanity project."

Axelrod's comment particularly annoyed me. I wouldn't have minded if he'd disagreed with my strategy or my assessment of the threat posed by Trump. I didn't even care if he thought I was crazy—it wouldn't be the first time someone thought that. What bothered me was that he claimed to know why I was doing what I was doing—and by going after my motives, he was trying to discredit our work. I saw that as an ad hominem argument.

More than that, I saw it as being cynical about a true threat to America. Democrats had spent years warning that if Trump were elected, he would be an existential threat to America. Events were proving them right. The logic was straightforward: if a president is behaving lawlessly, and putting the future of America at risk, then in our constitutional system, Congress has an obligation to try to remove that president from office, even if that carries political risk. Yet many leaders were opposed to both the risk and the way of thinking.

Whether you agree or disagree with me about Need to Impeach is water under the bridge. The broader point is that not acting with the courage of your convictions can be dangerous. Especially in the face of a danger the size of climate change.

One example of people taking the right theoretical position on climate, while being unwilling to act on their convictions, involves the unions that represent the building trades. When you talk to union leaders, they say they're taking climate seriously, and many of them back up those words with actions. The Service Employees International Union, for example, under the leadership of Mary Kay Henry, has shown that fighting tooth and nail for workers can and must include supporting the transition to cleaner, safer, healthier energy.

But other big unions act in contradictory ways. The International Brotherhood of Electrical Workers said in a 2021 policy brief for Congress that "the climate crisis is urgent and poses a threat to our nation's long-term prosperity." After President Biden announced his changes to the EPA in 2023, the United Auto Workers claimed in a press release that they supported "the transition to a clean auto industry." The United Steel Workers was a founding member of the BlueGreen Alliance, an organization created for labor and environmental groups to fight climate change together.

Yet when it comes to climate, these organizations too often struggle to fully walk the walk. While they recognize the threat, and tend to support elected officials who would address it, they frequently align themselves with the fossil fuel industry, giving a working-class veneer to those companies' billions in profits. The International Brotherhood of Electrical

Workers has publicly argued for continuing to use not just natural gas, but coal as well. The United Steel Workers filed a 2022 brief in favor of the Line 5 pipeline, which transports 22 million gallons of natural gas and oil every day. Leaders from the United Auto Workers refused to attend Biden's announcement of new rules designed to boost EV sales and productions. In fact, the UAW refused to endorse Biden for re-election over his commitment to EVs, even though he helped save their industry as vice president after the 2008 crisis and has arguably been the most pro-labor president in recent history.

I'm a big believer in labor unions. I've worked with union leaders, partnered with them on campaigns, and supported their organizing efforts. I recognize that first and foremost, unions exist to fight for their members, and I also recognize that for the better part of a century, fossil fuels supported a lot of union jobs, from work on oil rigs to manufacturing cars that run on gasoline. I can also see how, facing shrinking memberships, some unions might be wary of big changes and feel that they can't afford to act on climate.

But even given all that, slowing the transition to cleaner energy isn't in union members' interests. If the UAW hamstrings the American electric-vehicle industry, for example, it's not going to prevent electric vehicles from getting built and sold around the world. It just means they won't be made in America. Also, union members live on this planet like the rest of us. For trade organizations to truly stand up for their members, they need to think bigger and start coming up with ways to fight for workers' economic *and* physical well-being.

On top of that, the transition to clean energy can, if managed correctly, be a huge boost to the labor movement and the workers it represents. To bring down emissions, we'll have to electrify things like cars that run on fossil fuels, and power them with electricity from renewable energy, which we have to connect to the grid. You know who's going to get a huge percentage of those jobs? Union electricians! Meanwhile,

ArcelorMittal, one of the world's largest steel companies, estimates that each megawatt of solar power will require roughly forty tons of steel, while each megawatt of wind power will require 150 tons. That's a lot of potential jobs for union steelworkers.

Obviously, the transition to clean energy won't happen overnight. More important, the environmental movement needs to meet the unions where they are, make clear the benefits of working together, and fight to make sure that a fair share of clean-energy jobs are union ones. Union leaders need to know that climate activists have their backs. But if done correctly, taking the right-sized swing on climate is a chance for unions to secure well-paying jobs for their members for the next century. It's a way to increase their memberships and their power. And let's not forget that union members, like us all, will be a lot happier living on a planet that isn't defined by climate catastrophe.

Which brings me to a final important point about acting commensurately to the size of the problem: sooner or later, that's what happens by default. But the difference between acting sooner and acting later is enormous.

Climate people, myself included, often use a shorthand to describe the urgency of the moment, some version of "We have to act before it's too late." In many ways, it's a good way to get at the scale and size of the threat we face—if we don't get to net zero in time, we are likely to become trapped in a self-reinforcing feedback loop of runaway climate change that races out of our control. The problem, however, is that "act before it's too late" implies a binary outcome: we either do something in time; or we wait too long and do nothing, because there's nothing we can do.

If you're an executive at an oil and gas company, you might find this either/or choice very enticing. It might even suggest a strategy. If there's a point at which it's "too late," and transitioning away from fossil fuels is useless, the oil and gas industry has a strong incentive to delay until we reach that point.

But I don't think that's how things will play out. If climate stays on its current trajectory and goes from a series of discrete crises to a constant planet-wide catastrophe, people are going to take decisive, aggressive action. It's just that the action they take will be less predictable and, out of necessity, more disruptive than our efforts today. If rising temperatures lead to more droughts, farmers in South America who can't grow crops to feed their families won't sit around and politely starve to death—they'll cross any border they need to in order to save their kids. If rising ocean levels swallow up entire islands in the Pacific, the people living in those countries won't wait until the sea covers them—they'll go wherever they can to start a new life.

The Institute for Economics and Peace, a global think tank, has estimated that there could be as many as 1.2 billion climate refugees by 2050 if the planet keeps warming at its current rate. That's the equivalent of nearly the entire population of China needing to find a new home—in a world with less land people can safely live on, less farmland available to grow food to feed them, less clean water for them to drink, and fewer industries to employ them. There would be more poverty, more disease, and wars over resources. Entire regions would become destabilized. It would be devastating. The real choice isn't between acting now and acting never. It's between taking smart, constructive action right away that preserves our way of life, or opening a Pandora's box with billions of people taking desperate measures simultaneously and hoping it turns out for the best.

You probably don't have the ability to reshape global philanthropy or change the bureaucratic processes at the United Nations or spur the American labor movement to commit fully to the better future within reach. But as you do your own part on climate, you can ask yourself, often and honestly, "Am I doing enough?"

If you're reading this book, you're probably aware of the threat climate change poses. You probably are doing some things, maybe many things, to meet that threat. But if you haven't yet, this is the time to make your actions commensurate with the size of the problem, to unshackle yourself from the societally approved range of reason and base your decisions on what you know to be true.

You don't have to do everything. But you do have to do enough. And for most of us, that means doing a whole lot more.

CLIMATE PEOPLE

Leah Stokes

Every day, researchers and academics are conducting breakthrough climate studies, improving humanity's understanding of what is happening to our planet, what can be done to protect it, and how people's views on climate are changing. This information has the potential to shape public policy in ways that bring us closer to net-zero emissions. But distilling large volumes of research and getting it to elected officials with the power to change climate policy has never been easy.

That's where Leah Stokes comes in.

Leah's summers spent canoeing and camping in Canada gave her an early appreciation for the environment. In elementary school, she gathered milk cartons from her classmates for recycling. In high school, she wrote to her local grocery store, asking them to stop selling Patagonian toothfish because the species was threatened. By the time she got to college, Leah was running a campaign encouraging people to turn off their lights and conserve energy.

The campaign worked, with campus buildings decreasing energy use by roughly 12 percent. But Leah still felt that something was missing. "I just walked away feeling like this isn't big enough," she says. So, despite being a double major in psychology and East Asian Studies, she went into climate, spending the next decade working for environmental organizations, advising Canadian parliament, and ultimately getting a PhD in public policy.

Leah's research has focused on US energy policy, including the renewables movement, understanding opposition to wind energy and electrification, and factors that influence whether clean energy policies are likely to be enacted, expanded, or restricted. Her articles in journals, newspapers, and magazines across the country have increased climate awareness and have helped grassroots environmental groups more effectively persuade people to support climate action.

Her cumulative work helped shift the momentum on climate, and when politicians were finally ready to act, she helped them do that, too. After the 2020 election, she began advising Senate Democrats on climate policy, ultimately helping to write the Inflation Reduction Act—the largest climate legislation in American history.

The IRA will invest hundreds of billions of dollars in renewable energy, make it cheaper for people to install solar panels and electrify their homes, help develop new EVs, support climate-friendly agriculture, strengthen the Clean Air Act to reduce pollution, and protect wetlands and coastlands. Early estimates suggest it could reduce US emissions 40 percent by 2030 while preventing over 100,000 asthma attacks each year—and it wouldn't have been possible without Leah's decision to commit herself to making the biggest difference possible on climate.

"Her work has been instrumental in gathering people from the environmental movement, academia, and the clean-energy sector to work together for change," says Minnesota senator Tina Smith. "We'll experience the results of her remarkable leadership for decades to come."

CHAPTER IX

KINDNESS DOESN'T SCALE

t's a total cliché, but in my case it's true. The kindest person I've ever met was my mom.

Marnie Steyer liked people and could connect with anyone, no matter their background. She cared about people who were members of her family, and she cared about people who were wildly different from her. Most of all, she had a strong sense of self-acceptance that made her incredibly genuine. Like many women of her generation, she left the workforce to become a full-time mom, but once I went to school, she pursued a new calling: tutoring lower-income kids. Over decades, she made a difference in hundreds of children's lives. Putting others' interests before her own came naturally to her.

I honestly believe that if everyone was half as thoughtful, empathetic, and unselfish as my mother, there would be absolutely no need for this book. There would be no need for climate activism at all. Our species would have solved the problem decades ago.

But that's not how life works. I imagine that what's true for me is true for you, too. Most people aren't half as kind as the kindest person you know. Some people aren't kind at all. Even those of us who try to be decent toward our fellow human beings are, let's face it, often driven by our own self-interests.

Which is why I'm skeptical of any solution to climate that requires humanity's collective heart to grow three sizes. In a previous chapter, I talked about collective action. Collective action isn't just billions of individual good decisions stacked together. It's something else entirely. It's change at scale. Human kindness is a wonderful thing, but on an issue as complex, and as rife with self-interest, as transitioning away from fossil fuels, kindness alone is not a force powerful enough to transform the world.

What we need is a tool not just for creating change, but for convincing people around the world to embrace that change as quickly as possible. As it turns out, we already have one.

It's called capitalism.

Don't get me wrong. I'm not one of those people for whom capitalism is a kind of religion. I don't "believe in free markets" in some kind of ideological or spiritual sense. I don't think markets will inevitably make the world better. But I'm still a proud and committed capitalist. That's because I've seen firsthand that markets are—for better *and* for worse— the quickest way for human beings to change things at scale.

I suspect my views on capitalism are also shaped by the era in which I was born. Growing up in the 1960s, I saw firsthand what the American economy looks like when it's working the way it's supposed to. I don't mean that our market-based system was perfect or fair—it wasn't even close to those things, especially for women and Black and brown people. But the industrial giant that ramped up during World War II and saved the free world went on to provide a foundation for growing prosperity at home.

And not just for a handful of people, either. As our economy grew, the American middle class grew with it, and tens of millions of families were able to lift themselves out of poverty. For many of the people who lived through that time, capitalism felt like a miracle. Meanwhile, the horror stories coming out of Communist countries—famines, wars, dictatorships, gulags full of political prisoners—underscored that the most common alternative to capitalism was not utopia but its opposite.

This helps explain why, for a long time, I didn't have any concerns about my role in the machinery of markets. If someone had asked me, I probably would have said that I was part of a system that created more prosperity for everybody, and that the rising tide would lift all boats. How would it do that? I had no idea. I didn't think too hard about it. That's why I was able to love working out of Exxon's Alaska headquarters.

My views on the raw power of capitalism were also shaped by experiences I had at the start of my investing career. I saw firsthand capitalism's ability to bring down costs and send new ideas, technologies, and ways of doing business rocketing around the world.

A lot of the large-scale, market-driven change I witnessed involved oil. When I worked in Morgan Stanley's mergers and acquisitions department, our job was to give businesses advice on whether and how to buy one another. This was in the late 1970s, when the economy as a whole was in a slump but the price of oil was surging. As a result, at least half of the action, where M&A was concerned, was in the energy sector. In less than five years, the eighteen leading petroleum companies made eighty-five acquisitions, each valued at $15 million or more. By January 1984, the top fourteen of these companies had combined assets of $307.1 billion—almost $900 billion in today's dollars.

In theory, this was a textbook example of the free market solving a problem. Oil was a risky and expensive business, and those expenses were passed down to consumers. Consolidation lessened the risk, since the more drilling rights you owned, the less concerned you had to be that any one parcel of land would pan out. It also brought down expenses by making things like distribution and storage more efficient. This, in turn, drove down energy prices for everyone.

But even back then, for all my genuine faith that capitalism was improving most people's lives, I could see that markets are amoral. Markets don't care about making life better for people. They care about making things bigger. And the people who worked within markets almost always acted in their own interests. For example, here are two facts about my time at Morgan Stanley: first, we only got our fee if a merger went through; second, in almost every instance, we advised companies to go ahead with mergers even when they were buying something for far more than it was worth. As a young guy just out of college in his first finance job, I don't remember questioning our advice, but within a few months of starting I was surprised by how often we gave our clients what seemed like terrible advice. We weren't lying to them, but I don't think it was a coincidence that our suggestions almost always lined up with the bank's financial interests.

Our clients never complained. They shared the same short-term incentives we did. We never needed to go to a big oil company and say, "Hey, I really think you should buy some little oil companies." The people who ran big oil companies were always trying to buy little oil companies, because that would make their big oil companies even bigger. The bigger your oil company, the more valuable it becomes. The more valuable it becomes, the more you get paid. The more you get paid, the bigger a deal you are at the Petroleum Club of Houston. We ran the numbers and put together binders that gave oil companies cover for doing their deals. Then we arranged the financing. Often, consolidation made little sense over the long term, but that didn't matter. Everyone was making more money in the short term, which meant everyone was happy.

There was no nefarious plot to create huge oil companies; it was just capitalism doing what capitalism does. Companies want to become bigger companies. Most people want to become richer and more important tomorrow than they are today. I don't think that's a good thing or a bad thing. It's just a fact of life.

When I started my own business, I again saw the power of markets to scale things quickly. In investing, there are two ways you grow. First, if you do a good job, people give you more of their money to invest. The second way is through compound interest, which is like magic. If you get a 10 percent return on an investment, you can turn ten thousand dollars into eleven thousand dollars in one year. Not bad. But what if you keep doing that, year after year, for three decades? You end up with nearly $175,000—a more than 1,700 percent return on your investment.

Like many businesses, Farallon grew slowly at first, and then it grew really fast. We started with a guarantee that I would have $6 million to invest, which was, even then, a tiny amount by investment-firm standards. By the end of the first year, we had about $40 million. By the time I left, we had more than $20 *billion* in assets under management—a growth rate of more than 87,000 percent over twenty-five years.

Theoretically, Farallon's growth benefited regular people: to use just one example, when we invested for pension funds and universities, the return on that investment meant more money for retirees or cutting-edge research. But that's not *why* we grew. We grew because in a free-market system, businesses are always trying to grow. If the way to make your company bigger is to distribute more lifesaving vaccines or bring clean drinking water to communities that don't have it, then growing your company is good for the world. If the way to make your company bigger is to distribute more cigarettes to teenagers or dump chemicals into streams, then it's bad. But again, markets don't care. They're amoral: they operate in the world of numbers, not values.

Ironically, the better I did in the markets personally, the more skeptical of unrestrained capitalism I grew. This was partly because of Reaganomics. Rich people got richer than ever, but the rising tide lifted all boats far less effectively than it had before. I also saw that capitalism, despite or perhaps because of its raw power, gave people overwhelming incentives to follow their self-interest at the expense of everything else.

One example I remember vividly involved Monsanto, the chemical giant that makes effective—and highly toxic—chemicals for agriculture. As was already becoming clear at the time, and has since been affirmed in court, Monsanto makes products that are closely linked to rising rates of cancer, especially in the places where those products are manufactured. One day about fifteen years ago, a close friend and very successful investor, someone I'd always considered a good and decent guy, said, "I really like Monsanto as a stock."

I discussed it with Kat, who was already deeply involved in issues surrounding agriculture. If anything, she was even more concerned than I was. I went back to my friend.

"Okay," I said, "but it seems to me that these guys are kind of, you know, poisoning people. Do you really want to be involved with that?"

He looked at me like I was nuts. In a way, I was. He was very good at investing, which meant that his analysis about the stock was probably correct—as far as it went. And because of the incentive structure in investing—the better your returns, the higher your compensation—I was basically asking him to take a pay cut in exchange for doing the right thing. In the end, he dismissed my concerns.

But the story doesn't end there. Years later, when I was running for president, I spent a lot of time in Iowa. I learned that Iowa has the second-highest cancer rate in the country, and it's because of the agricultural runoff. Inevitably, all the toxic insecticides and fertilizers farmers put on crops find their way into the water supply, and people drink the water, or their skin comes into contact with chemicals, and they get cancer. On one of my visits to the state, I went to the farm of a very successful industrial farmer who used all the latest chemicals, and the very last thing he said to me was, "I have cancer." He was just sixty-two years old. Not long after that, I was talking with my former partner—the one who'd thought Monsanto was a good buy—and I told him this story.

"What kind of cancer?"

"Non-Hodgkin's lymphoma."

"Oh, that's no big deal, it's treatable. What's your point?"

Eighteen months later, the farmer died.

I can't think of a better warning about the dangers of unrestrained capitalism. Sometimes markets create an information revolution or pull families out of dire poverty. Other times they can lead someone to dismiss the plight of a farmer with cancer.

Or to expand the fossil fuel industry.

For decades, the incentives to extract oil and gas were the same as those that compelled my colleague to buy Monsanto. People loaned money for new drilling projects, or founded start-ups to make the fossil fuel industry more efficient, or advised autocratic governments of petrostates, or took jobs at oil and gas companies—not because they believed in fossil fuels in an ideological sense, but because oil and gas made tons and tons of money, and getting your hands on some of that money was a very effective way to buy a bigger boat. In a world powered by human kindness, clean energy would have replaced fossil fuels a long time ago. But in our world, in which large-scale trends are driven by markets, fossil fuels had a clear advantage for a very long time.

That's changing fast. In many cases, it's already changed. And as climate people, we need to make sure that the tables turn even faster. How can we shift the incentives so that even the most self-interested people make choices that help stabilize our planet and prevent human catastrophe? What does climate capitalism look like?

The way I see it, we'll need four things: better tech, better ideas, better rules, and better metrics. Each of the final chapters of this book will be devoted to one of these necessities.

Let's start with tech.

For a long time, a lot of people in the climate movement felt that the key to succeeding in the marketplace was to get people to pay a so-called "green premium." There were two related versions of this idea. The first was that people would pay extra for products that are good for the planet because they want to demonstrate their commitment to the planet, the same way someone might pay extra for cage-free eggs because they care about animal welfare. The second was that because burning fossil fuels comes with enormous long-term costs—including what economists call "externalities" such as disaster relief or damage to public health—we should be willing to pay higher upfront prices for clean energy and green products because they come at a lower cost to society over the long run.

I fundamentally disagree with the premise behind both these versions of a green premium, for two reasons. First, in a competitive world, selling more expensive stuff doesn't work. Some people might have the interest, and the disposable income, to pay extra for products because they care about the planet and want to demonstrate that commitment through their purchases. But that's another version of kindness, and as admirable as it may be, it won't scale. Getting to net zero will require transitioning the entire world away from fossil fuels, and that means making clean energy and cleantech the least expensive option. Along the same lines, externalities are real—economists should continue to measure them, and policymakers should take them into account. But if we want a real shot at stabilizing at our planet, we have to compete on sticker price.

My second problem with the idea of the green premium is that it implies that such competition isn't possible, that cleaner automatically means more expensive. That's not true—and it's becoming less true every day. Just look at the American coal industry. While policy has at times had an impact, and advocacy, particularly the Sierra Club's "Beyond Coal" campaign, has accelerated existing trends, the real reason we haven't built a new coal-fired power plant in years and that the number of coal miners has shrunk from over 215,000 in the 1980s to fewer than 40,000 today is pretty simple. Coal got too expensive. Cleaner energy won in the marketplace.

Unfortunately, that cleaner energy was still pretty dirty—for the most part, we transitioned from coal to natural gas, which (due to the methane leaks I mentioned earlier) is nearly as bad, and sometimes even worse, than coal. But what natural gas did to coal, renewable energy can do to oil and gas. In fact, in many places, the cheapest form of electricity is now solar or wind. Many of the cheapest, best vehicles are electric, too. And that's not even taking into account the massive taxpayer subsidies that the oil and gas industry receives.

Back in the 1800s, Ralph Waldo Emerson wrote that if you can build a better mousetrap than your neighbors, the world will beat a path to your door. The same idea applies when it comes to energy, and things that use energy. If we can build better, cheaper stuff, people will buy it. And if that stuff happens to protect the planet, improve the quality of life for everyone, and prevent hundreds of millions of needless deaths, we can harness the power of capitalism for good.

That thinking is what led Kat and me, almost immediately after the Alaska trip where we saw the vast empty space where a glacier had been, to Stanford University. For decades, Stanford has been on the cutting edge of tech, not just funding groundbreaking research, but turning that research into technology and products that succeed in the marketplace. Google, Yahoo, and Snapchat are just a few companies that began there. The IT revolution scaled because of market capitalism, but the seeds were planted at Stanford University.

We asked the president of Stanford, John Hennessy, "Why can't we do for cleantech what Stanford did for IT?" We then gave Stanford a $40 million grant to create a program to fund innovative ideas. Because it was a grant, not an investment, we knew we would never see a dime from any companies that emerged from the research of the grad students and professors we supported. But today, that $40 million grant has led to businesses with over $6 *billion* in value. Here's just one example:

Today, in the United States alone, there are about three million tractor trailers that run on diesel, not to mention millions more small- and medium-duty trucks. Diesel fuel consumption accounts for about 25 percent of the overall carbon pollution from the transportation sector. But electrifying trucks is difficult, because the batteries required to power them would have to be really big (at the moment, anyway). In 2016, fellow doctoral students BJ Johnson and Julie Blumreiter came up with a solution. Instead of trying to replace diesel engines entirely, their company, ClearFlame Engine Technologies, modifies existing diesel engines so that they can use renewable fuels like ethanol and methanol. A ClearFlame engine keeps 80 to 90 percent of the components from the original engine, and trucks that use their engines can cut their carbon emissions almost in half. It's a win for truck owners and drivers, too, because ethanol is cheaper than diesel.

Over the last few years, through Galvanize, Katie Hall, my partner at Galvanize, and I have also invested in plenty of companies that didn't have their origins in Stanford research. One of them, Alcemy, aims to revolutionize the way we make concrete. Concrete is a huge—and often overlooked—source of carbon pollution. This is a particularly challenging issue because countries that are lifting their people out of poverty tend to build a lot of buildings very rapidly, and thus use a lot of concrete. It had long seemed that we would have to choose between prioritizing a livable planet or prioritizing livable buildings in developing countries. Alcemy wants to change that. The founders, Leopold Spenner and Dr. Robert Meyer, use machine learning and control technology to make predictions about cement mixtures and their ingredients. Its software allows concrete producers to lower their carbon content while maintaining, and in some cases improving, the quality of the finished material.

Of course, for every company I've helped support in some way, there are countless others I've had nothing to do with. I'm just glad they exist. For example, in the 1980s, Donnel Baird's family immigrated to the United States from Guyana, moving into a small one-bedroom apartment in Brooklyn without functioning heat. In the winter, his parents would turn on the kitchen stove and oven for warmth, opening the windows to let out the carbon monoxide. It's hardly surprising that Donnel became keenly interested in the science and business of heating and cooling buildings.

But when Donnel joined the Obama Administration to help run a multi-billion-dollar effort to retrofit old structures, he concluded that the program, while well-intentioned, didn't work. The problem was that every building was different. It took a lot of time and money to figure out what changes were needed, and even with federal subsidies, landlords balked at the cost. So he started a company, BlocPower, which uses custom-built software to inexpensively create a digital model of nearly any building and come up with a customized solution. With a focus on buildings in low-income and underserved neighborhoods like the one Donnel grew up in, the company installs all-electric heat pumps to replace the ones that burn oil and gas. Best of all, BlocPower pays for the entire upfront cost of the investment. Building owners pay back Donnel's company over time, but because their utility bills are so much lower with the new heating systems, they still save money each month.

That's what winning in the marketplace looks like. Not just cleaner. Cheaper and better, too. If you own a truck fleet, a concrete mixer, or a residential building in New York City, it doesn't matter if you care about climate change or not. Upgrading to cleantech is just business common sense. To put it another way, if you want to stick with fossil fuels, it will cost you more than switching to something cleaner.

That's one reason the green premium idea is misleading. More and more often, companies and consumers can cut emissions without paying extra. In fact, in a growing number of cases, they'll have to pay a polluters' premium to stick with fossil fuels.

The fossil fuel industry knows that the tables are turning. Increasingly, products that are helping us stabilize the planet aren't just better for the public at large than their oil-and-gas-powered counterparts—they're cheaper and better for individual customers, too.

Which is exactly why fossil fuel companies and their allies are trying to mislead the public now—before the adoption of clean-energy-powered products becomes widespread enough that everyone discovers they like them.

In 2023, for example, the *National Review* published a big story called "The War on Things That Work." The author, Noah Rothman, warned that scary environmentalists want to take away all your household appliances, from your gas stove to your air conditioner to your lawn mower, and replace them with electric junk that will break. "Armed with unchecked self-confidence and possessed of an abiding faith in the idea that you must be coerced into altruism," he wrote, "the activists seem to be coming for almost everything you own." This is total nonsense, but when people associated with the climate movement spend all their time talking about how much more expensive saving the planet is than burning fossil fuels, it makes the war-on-things-that-work argument seem much more convincing than it should be.

In fairness to Mr. Rothman and the *National Review,* not so long ago the environmental movement was mostly asking people to make sacrifices: drive less; fly less; install expensive solar panels. For some people, knowing that their travel or hot water or electricity wasn't harming the planet was worth the extra cost. But for most people, it wasn't. People wanted to buy things that were cheaper, better, and ideally both. For many, that meant buying oil and gas—or products that needed oil and gas to work.

But the *National Review*'s spin ignores what is, for them, an inconvenient truth. "Green" technology and products are no longer pie-in-the-sky ideas. People aren't buying them because they're liberals. People are buying them because they like getting more bang for their buck. Or to put it in the *National Review*'s terms, even without the massive, nonstop bailout the oil industry receives when taxpayers pay to clean up its mess, the "things that work" happen to be the things that work for the planet.

One of the most obvious, widespread examples of this involves solar panels. In 1979, when Jimmy Carter became president, he put solar panels on the roof of the White House. It was a virtuous gesture—the panels were clunky and expensive to install, and any electricity they generated would have been wildly expensive. At the time, solar energy cost roughly fifty times as much as energy generated by fossil fuels. When Ronald Reagan took office, he ripped out the solar panels as a demonstration of his commitment to putting free-market economics over bleeding-heart lefty values.

Today, millions of American homes, including the White House, rely, in whole or in part, on solar. The cost of solar panels has fallen by 99 percent since 1977. Rooftop solar isn't just cleaner than traditional power, it's far cheaper. Some solar companies have made solar installation basically free, by using the money homeowners save on cheap solar energy to pay for the panels themselves. In other cases, homeowners are actually making money via solar, by selling energy back to the power grid. On average, solar is now 33 percent cheaper than natural gas.

That gap in price is almost certain to keep growing, because prices for new technologies tend to go down much faster than prices for things that have been around forever. (Think of what's happened to the cost of a gigabyte of computer memory over the last two decades versus what's happened to the cost of a book. The former has gotten more than one thousand times cheaper; the latter has gotten more expensive.) The price of oil shoots up every time the cartel that includes Russia, Saudi Arabia, Iran, and other petrostates decides to artificially cut supply. Meanwhile, solar just keeps getting cheaper. Batteries continue to improve, too, making solar more reliable and increasing the amount of energy that can be stored and sold.

If Ronald Reagan wanted to champion free markets today, he wouldn't be ripping out solar panels—he'd be putting them in. It's the fossil fuel people, not the climate people, who are asking customers to pay a premium to support their political agenda.

Now imagine the same trends that have already transformed the market for household energy transforming the market for everything. That's what's happening right now. In some cases, the economic equation hasn't yet flipped—for example, green hydrogen, a zero-emissions fuel that could power trucks, cargo ships, factories, and even airplanes, is still more expensive than fossil fuel. But we can see where things are headed. In a large number of other cases, unless you happen to love the fossil fuel industry and want to increase oil and gas companies' profits, there's no reason to keep using obsolete technology. It costs you, the customer, less to buy an electric vehicle than a gas-powered car, and it's much, much cheaper to maintain one, since you have far fewer moving parts that can break. It's cheaper to heat your water with electricity than with natural gas. It's cheaper—not to mention more convenient, quieter, and safer—to run your lawnmower or power tools off a battery instead of gasoline.

Yet for all the promise of the cleantech revolution, I want to sound a cautionary note: innovation alone is not enough to win in the marketplace. A viable business is very different from breakthrough tech.

A big mistake that people make throughout the private sector is thinking that success is all about the product. They imagine a straightforward competition: "Jill's tech is a 93 out of 100, and Jack's tech is an 87, so Jill wins."

That's not how it works. A friend of mine who started an electric vehicle company around 2010 insists that in terms of building batteries and other tech basics, Tesla is only now getting to the place his team reached fifteen years ago. But it doesn't matter, because my friend's company is defunct. He couldn't raise the money to keep going, and Tesla, for whatever reason, could. The ability to win people over—to have a vision, to sell, to keep going through tough times—is every bit as important for succeeding in business as having a low-cost, viable product. Which is all the more reason for people from all walks of life to join the climate movement. If you're a brilliant researcher who can summon technological breakthroughs seemingly from nowhere, it's probably not too difficult to figure out how you can become a climate person. But if you've worked at any business at all, we need you, too. I know from my own experience that starting one's career outside the climate world can be a huge asset when you're ready to join the fight and help stabilize our planet.

Climate people aren't the ones asking their neighbors to sacrifice for the sake of the greater good. We're the ones creating incredible new businesses that help our neighbors get better stuff for less money. It's the fossil fuel industry—and its hangers-on in the media—who, for purely ideological reasons, are desperate to make people pay for products that are more expensive, less reliable, or both.

Remind me again who's waging a war on things that work?

CLIMATE PEOPLE

Olivia Dippo and Andy Zhao

We don't always think about it, or even see it, but steel is everywhere. We use it to build bridges and reinforce buildings, to make refrigerators and washing machines, even to construct the wind turbines that generate renewable energy. Steel is one of the indispensable materials of modern life.

But there's a problem. Making steel requires huge amounts of fossil fuels. The process, which has gone unchanged for centuries, typically goes something like this: iron ore is mined; the impurities within it are extracted in blast furnaces that are typically heated by burning coal; and a small amount of carbon is added to the liquid iron, again using blast furnaces. Once allowed to cool, the resulting material is steel, and each step involved in its creation releases huge amounts of greenhouse gas into the atmosphere. In fact, steel production accounts for roughly 8 percent of global greenhouse gas emissions.

Going without steel isn't really an option in modern society, so the oil and gas industry would love for us to think that all this carbon pollution is an unavoidable cost of life. But what if it wasn't?

That idea, and the mission behind it, is what inspires Olivia Dippo and Andy Zhao, two recent PhD graduates from the University of California at San Diego. As a graduate student at Carnegie Mellon University in Pittsburgh, part of Olivia's work focused on 3D printing and how light interacts with materials, specifically how it can melt metals. Andy, meanwhile, had long been interested in clean energy and had worked on ways to make solar and nuclear energy more efficient. Over weekly coffees, Olivia and Andy discussed whether the technology behind 3D printing could support renewable energy.

Eventually, they came up with an idea that sounds like something out of science fiction: making steel with lasers.

To put their theory into practice, they moved to Oakland and started a company, Limelight Steel. Instead of using fossil fuels to heat up iron ore in a blast furnace, Limelight uses light energy from lasers. Iron ore's natural properties allow it to absorb this light as heat—it's similar to the way microwaves can heat water using targeted energy waves. By comparison, a traditional electric-powered furnace operates more like a giant toaster oven, with coils that take time and significant power to heat up. The efficiency of lasers allows Limelight's system to produce steel using as much as 45 percent less energy than is typically required, and to do so running entirely on electricity. If that electricity comes from renewable sources, net-zero-emissions steel is within our grasp.

Olivia and Andy believe that as their technology develops, they'll be able to disrupt a two-hundred-year-old industry. That will allow us to build everything from bridges and airports to cars and kitchen appliances—while cutting carbon pollution by billions of tons per year.

CHAPTER X

RULES MATTER

n 1962, Rachel Carson, an aquatic biologist and science writer, published *Silent Spring*. In it, she showed that pesticides—and in particular the chemical known as DDT—were having devastating effects on the natural world and poisoning human beings who were unknowingly exposed to them. What's more, she exposed the chemical companies who knew about DDT's horrible effects but either flat-out lied or withheld that information from the public. *Silent Spring* is one of the most influential books written in my lifetime. Many people say it launched the modern environmental movement. At the time, it received near-universal acclaim for its bravery and poetic style.

But not everyone was excited about *Silent Spring*, and many critics decided they knew Carson's secret motives for writing it. "Miss Rachel Carson's reference to the selfishness of insecticide manufacturers probably reflects her Communist sympathies," declared one letter to the editor of the *New Yorker*, which had published excerpts of Carson's book. Ezra Taft Benson, a farmer and former secretary of agriculture, wrote Dwight D. Eisenhower to say that Carson was "probably a Communist."

The chemical industry mounted a similar smear campaign. Journalist William Souder recounted in *Slate* one pesticide maker's attempt to discredit *Silent Spring*: "Carson, the company claimed, was in league with 'sinister parties' in order to further the interests of the Soviet Union and its Eastern European satellites."

Other critics refrained from attacking Carson personally but warned that if, motivated by public outcry, the government were to rein in pesticides, it would threaten the American way of life. Velsicol, a chemical company, wrote Carson's publisher to warn that "our supply of food will be reduced to East-curtain parity" if new environmental rules were put in place.

You probably know what happened next. The United States banned DDT for agricultural use in 1972. Somehow, despite this, our food supply remained the envy of the world. We even won the Cold War.

Many of the most harmful pesticides Rachel Carson warned us about are now gone. But remarkably, the basic argument made by the chemical companies in 1962 lives on today. Calling themselves free-market champions, calling their detractors communists, and suggesting that new regulations will cause the free market to collapse are among the favorite talking points of the fossil fuel industry. The right-wing Heritage Foundation called the Inflation Reduction Act—President Biden's massive investment in clean energy, which mostly flows through private companies—"Big Government Socialism." Not to be outdone, Larry Kudlow, a conservative pundit and the former chair of Donald Trump's National Economic Council, called it the "ultimate big government socialism bill." Billionaire publisher and former Republican presidential candidate Steve Forbes called the push to increase electric-vehicle sales "the unsustainable folly of modern socialism." You get the idea.

It's hard to list all the ways this talking point is just plain wrong, but to me, it starts with an awareness that there's no such thing as a completely free market. As I well know from my years of investing, every business must comply with a huge number of rules and regulations, both large and small. But just because companies are required to make certain information available to the SEC, or to meet basic workplace safety standards, or are forbidden from engaging in insider trading doesn't mean we're living in some kind of communist dystopia. It just means that in business, as with anything in life, there are rules.

Not every rule is perfect, of course. There are rules that don't make sense or have the opposite effect from the one intended. But broadly speaking, we understand that rules are necessary. No one thinks you should be allowed to make money by holding other people up at gunpoint—or that making mugging illegal is somehow socialist.

When done right, rules reflect the values of the countries and communities in which businesses operate. They allow for innovation, and for the raw power of capitalism to create change quickly at scale. But they also help set a direction for that change, ensuring that, broadly speaking, the incentives of companies—and the people who run them—line up with the interests of the public at large. And there's no more urgent interest for our society at this moment than preventing a climate catastrophe. So the challenge facing climate people, particularly those in government, politics, and policy, is not whether to have rules. It's how to write better ones.

How can we make sure that the power of the market protects, rather than irreparably damages, our country, our people, and our world?

When rules work the way they're supposed to, they form a feedback loop with the other elements of climate capitalism. Better companies come up with breakthrough technologies and ways to scale them. Better ideas and messages encourage people to embrace those technologies and help policymakers understand why they're so urgent. Better rules create an environment where existing great companies can grow, and where new ones can flourish.

The oil and gas industry says that if we take a more sensible approach to rules, it will hurt our economy. I think it's the opposite. If we don't replace outdated rules with better ones—ones that help us lead the clean-tech revolution rather than force us to lag behind as countries like China pass us by—we'll never be able to unleash our economy's full potential.

Done right, protecting people doesn't stifle innovation. It spurs innovation. Not just when it comes to climate, either. Take what's happened with auto safety. The year before I was born, in 1956, the highway death toll was nearly 40,000. Given how many fewer people lived in the United States back then, it's the equivalent of 85,000 deaths today. More Americans were killed by automobiles that year than in battle during the entire Korean War.

For the last several decades, the fossil fuel companies have tried to blame their own customers for climate change. (Remember the BP carbon footprint calculator, designed to make you feel as though global warming was your fault rather than the inevitable consequence of a worldwide economic system that runs on oil and gas?) Back then, the car companies of the 1950s insisted that the fault for automobile fatalities lay entirely with drivers. A breakthrough technology—the seatbelt—was catching on, but it was viewed with skepticism by consumers. Some didn't understand how seatbelts worked. Others didn't like the way they looked or felt or didn't want to be reminded that they could get in an accident. The result was a vicious spiral: the car companies didn't have much incentive to develop new and better safety technology, so drivers weren't exposed to it, so they didn't see its benefits and the need to demand it, so car companies didn't have much incentive to develop it, and so on.

Then, in 1965, Ralph Nader shocked the country with a big idea. In his book *Unsafe at Any Speed*, he argued that the reason for the enormous number of driving fatalities wasn't bad drivers—it was neglect from the car companies. In 1968, the federal government required lap and shoulder belts in all new passenger cars sold in the United States. Suddenly, car companies had a good reason to invent better seatbelts—the modern car seatbelt, the one you almost certainly have in your car today, came along just five years later. And the new technologies were not limited to seatbelts. Padded dashboards, collapsible steering columns, reinforced fuel tanks, shatterproof glass, stronger brakes, and flexible bumpers all were introduced in the wake of federal safety regulations.

In 1971, in a tape-recorded White House meeting, Ford CEO Lee Iacocca complained to President Richard Nixon that safety rules "have really killed all our business." That wasn't true. Safety rules had *changed* Ford's business. Auto manufacturers didn't stop coming up with ways to make cars accelerate faster or handle more responsively or look cooler. But thanks to new rules about how the automobile market operated, they had to make cars safer, too, if they wanted to compete. They couldn't get ahead by cutting corners on safety—secretly killing customers no longer gave businesses a competitive advantage.

Better rules worked. In 1965, there were about five deaths for every 100 million miles driven in America. In 2022, the most recent year for which data is available, there were about 1.3 deaths for every 100 million miles. To put it differently, rules promoting safety didn't destroy the American auto industry. By changing the incentives for companies in the automobile market, they saved nearly half a million American lives and counting.

Those of us in California lived through an almost identical scenario—only this time, car companies were required to help protect their customers from air pollution rather than car crashes. When I first moved to California, the southern half of the state, and in particular Los Angeles, was still synonymous with smog. While the situation had improved since the passage of the Clean Air Act in the 1970s, pollution was not a theoretical problem. It was real, and it was hurting people. So in the early 1990s, state regulators came up with a rule to spur the development of cleaner cars. Eight percent of sales would need to come from vehicles that produced nearly zero emissions, and 2 percent of sales would need to come from vehicles that produced no emissions whatsoever.

Before the rule could go into effect, General Motors sued our state and, predictably, attacked the new rule as a danger to free enterprise. "You don't mandate markets," GM spokesman Chris Preuss told the *New York Times*, adding that the new rule was "completely unworkable." And GM was not alone in opposing the new rules—every car company was against them. "Nobody believes in it," said Carlos Ghosn, the chief executive of Nissan at the time. "But it's part of the cost of doing business."

Mr. Ghosn probably didn't realize it, but he was making an articulate case for better rules. It shouldn't be up to car companies to decide whether we get to breathe clean air or not. That's a choice We, the People, get to make. At the same time, we can't penalize companies who decide to do the right thing by giving an advantage to companies that cut corners. The car companies may not have liked the new California rule, but it put them all in the same boat.

Before long, despite all their complaints, car companies were innovating. In preparation for the new regulations, automakers invested in hybrid engine technology. They added catalytic converters to reduce smog. They created electric versions of their existing SUVs.

In fact, looking back, I suspect the companies' biggest regret is that they didn't embrace the new rule enthusiastically enough. One *New York Times* article reported, "Auto executives say that 2 percent requirement has forced them to keep alive a technology they would just as soon give up on: the battery-powered automobile." We can only imagine how well a major car company that went all-in on EVs in 2002 would be doing today, or how much further behind they would be if California's new rules hadn't gone into effect.

Today, new rules are focused on curbing carbon pollution. What amazes me, even though I've spent decades in business, is how we're repeating the same cycle—an industry insisting that the sky is falling, then getting to work and innovating new solutions—yet again. Often, the exact same companies are warning that the end of the free market is nigh. In 2012, for example, the Obama Administration put in place new miles-per-gallon standards. The new rule—which required automakers to sell cars that got an average of 54.5 miles per gallon by 2025—was projected to remove about 6 billion tons of carbon pollution. According to the *New York Times*, it was "the single largest federal policy ever enacted to reduce climate change."

Almost immediately, the automakers began to cast doubt on whether they'd be able to meet the new standards. So I called up a fellow Stanford business school grad named Thomas. Thomas understands markets and economics in general, and his family ran a GM dealership, so he also knows exactly how the car business works.

"GM keeps saying these mile-per-gallon standards are impossible to meet. Is that true?" I asked.

"No," he replied. "It's not true."

"Well, they're saying they can't do it."

"Yeah. Because they don't want to do it."

He explained to me that GM's strategy at the time was designed around making the car the home entertainment system for your family. That's where they were focusing their innovations: personal TV sets in the seats, sound systems, DVD players built into the dashboard, and so on. All that stuff takes up a lot of energy, and it's really heavy, so it would have been a lot more difficult under the new EPA rules. But plenty of other car designs were possible, as long as you didn't care as much about playing DVDs in surround sound as you drove.

In other words, the new rule didn't stifle innovation. It shifted innovation from one area to another. Engineers went from answering the question "How can I pack as many personal entertainment devices as possible into a car?" to "How can we use less gasoline and save our customers money without compromising the quality of the ride?" When I was a young father, I would have loved some TVs in the back seats to entertain my four kids. But I care a lot more about having a livable planet for my kids than I do about having a home-entertainment center for them on the road. Better rules spurred innovation, putting the scaling power of capitalism to work for the common good.

It's also worth noting what the Obama-era miles-per-gallon rules *didn't* do. Instead of the government saying, "Here's the exact set of fuel-saving technologies you need to install in every car," they said, "Here's an ambitious target everyone in the industry needs to hit, but it's up to you to figure out how to do it."

I think that's smart. More often than not, when it comes to climate rules, standards are better than mandates. Coming up with rules should involve a level of humility. We don't know exactly how the clean energy revolution is going to play out. We can't tell which promising technology might underdeliver on its promise, or which far-fetched idea might suddenly become not just possible but ubiquitous. Green hydrogen, for instance, is in theory one of the cheapest, most miraculous alternatives to oil and gas on the market. The problem is that producing green hydrogen costs between four and seven dollars per gallon, but to compete with oil and gas, we need to bring the price down to about one dollar per gallon. That's difficult, because transporting liquid hydrogen safely is, for the moment anyway, quite expensive. There's a chance green hydrogen will be the fuel of the future, but it's far from certain.

Which is why, instead of mandating that X percent of vehicles must run on green hydrogen by year Y, it's better to say, "We need to cut emissions by X percent by year Y." Let's allow the market to figure out how best to do it. If you believe the best path to reducing emissions is bringing down the price of green hydrogen, you'll work on that. If you think we're more likely to reach emissions goals by inventing longer-lasting batteries, or developing regenerative agriculture, or taking carbon out of the air, you'll work on that instead.

A fair standard creates healthy competition. If you can come up with a better way to help us cut emissions, your company will outpace its rivals and you'll make more money. Instead of the oil-and-gas race to the bottom, where people do better by making the rest of us worse off, we'll see a race to the top, where people pursuing their self-interest nevertheless end up doing something that's good for the planet and the public as a whole.

Even policymakers are often surprised by how effective rules can be. Many politicians, particularly among Democrats wary of being accused of working on the side of big government, are more comfortable with subsidies than with rules. In essence, they'd rather encourage companies to act in the public interest than prohibit them from acting against it. I get that. I'm a big believer in the importance of investment to fight climate change, and government investment can help complement funding that comes from the private sector. Also, there are times—as with the Inflation Reduction Act—when public investment is politically possible and rules are not.

But the simple truth is that rules work. Not just that—they often work far better than any other mechanism we have to help shape markets. History is full of examples where much-needed change was taking far too long to materialize, until new laws or regulations provided a jump start and improved people's quality of life. Perhaps the most striking example is London around the time of the Industrial Revolution, which became famous for having some of the most noxious air on Earth. Rising from millions of chimneys around the city, from both houses and factories alike, soot and sulfur dioxide mingled with fog to create what Londoners called "pea-soupers"—dense, deadly miasmas that hovered over the streets below, varying in color from yellow-green to brown to black.

London Fogs were not just unpleasant. They were deadly. Cars crashed into pedestrians in the darkness. Crime rates rose on foggy days. From

1840 to 1890, deaths from bronchitis alone rose 1,100 percent. Facing a full-fledged environmental disaster, Londoners organized. Societies dedicated to "smoke abatement" formed and raised awareness. In the late 1800s, one group even held an exhibition for smoke-related technologies, which attracted 100,000 people eager to learn more. But despite their best efforts, these appeals to the public interest and attempts to foster innovation were mostly unsuccessful. As residents left for the suburbs, the city's air quality improved somewhat, but pea-soupers remained a regular, noxious feature of London life. In 1952, the deadliest fog in history descended on the city, killing 12,000 people. There were so many funerals that London's florists ran out of flowers.

In response, the British parliament decided to do something that they had previously dragged their feet on: over the objections of the country's electricity-generation industry, which at the time relied primarily on coal, they passed tough new legislation. Britain's Clean Air Act, which went into effect in 1956, was the world's first comprehensive, national air pollution law. It outlawed the emission of "smoke nuisances" or "dark smoke" and required new furnaces to emit little or no smoke. It also offered grants to homeowners to replace their polluting appliances with electric ones, expediting the switch from coal to gas. In just a decade, London's smoke emissions declined 76 percent. Lives were saved, and Londoners still had the electricity they needed. Rules did what public awareness campaigns alone could not.

More than fifty years later, and thousands of miles away, my home state of California experienced something similar. We have a cap-and-trade system that creates a market for pollution. We're home to many of the most innovative companies, founders, and researchers in the world. These matter—I've already written about how important they've been to the clean-energy transition that is now underway.

But the most important thing California has done on climate is put rules in place. Our state's clean air and clean water acts go well beyond the federal ones. Our clean car rules have jump-started the adoption of EVs. And because we're the world's fifth-largest economy, changes that happen in California have a tremendous influence on the rest of the country. We've led America, and the world, not just in setting an example, but in setting higher standards that make it harder for companies to profit from destroying the planet. What we've seen, again and again, is that businesses might complain at first. But then they do what they do best: innovate, solve problems, and compete.

It always frustrates me when the fossil fuel industry acts as if businesses are unable to adapt to new rules in the market. Adapting is what we do best! It's how we won World Wars, and the Cold War, and built the strongest economy in the history of the world. Any time I hear someone say that requiring businesses to stop profiting from pollution would kill the economy, I want to ask that person, "What do you have against American innovation?"

Also, the idea that fossil fuel companies are opposed to writing new rules for their industry just isn't true. They're fine with more regulation—so long as that regulation benefits them.

In 2022, the last year data were available, oil and gas spent $124.4 million lobbying the federal government—and the amount is higher when you include the lobbying done in state governments. And that's just the money they're obligated to disclose. It doesn't include the dark money they're spending through SuperPACs or faux-grassroots groups that try to hold up clean-energy projects. The fossil fuel industry isn't spending all this money to advance the cause of free markets. Instead, they want policymakers to ensure that the rules of the markets favor oil and gas even more than they already do.

Today, for all the scaremongering over climate regulations, the legal playing field is tilted massively toward fossil fuels—often in ways the average American isn't aware of. You probably haven't heard about the "oil depletion allowance," but it's one of the many ways the fossil fuel industry has written the rules to get an unfair advantage in the market. It's technical and a little boring (a combination that lobbyists favor, since that means most people don't pay attention to it), but it basically works like this. When you drill for oil, you can deduct certain costs from your taxes. These costs include some you'd expect: the cost of discovering oil, purchasing drilling rights, the machinery to get your well up and running. But you can also deduct 15 percent of the gross revenue you make from selling the oil you drill.

Imagine if you could deduct 15 percent of your income from your taxes for no reason. That's what the oil companies get to do, every single year, and it adds hundreds of millions of dollars annually to their profits. From a tax-law perspective, investing in oil and gas is one of the most advantageous investments you can make in America today. In fact, that's the whole point of the tax break. President Eisenhower defended the oil depletion allowance by saying, "There must certainly be an incentive in this country if we are going to continue the exploration for gas and oil that is so important to our economy." The idea, back then, was that drilling for more oil in America was a good thing, so we should change the rules to create incentives within our market system to allow for more drilling.

Today, the national interest has changed—and public opinion has changed with it. According to a 2023 poll from the nonpartisan Pew Research Center, when people were asked to choose which energy priority was more important, "Developing Alternative Energy Sources" beat "Expanding Production of Fossil Fuels" by a thirty-five-point margin. Even among Republicans, 70 percent of those polled would like to see the United States increase its production of solar energy, and 60 percent would like to see more wind power. Yet solar farms and wind farms don't get to deduct 15 percent of their profits. Only the fossil fuel industry does. Oil and gas like to pretend they believe in free markets when in fact they believe in continuing to stack the deck against their competition.

Nor is the fossil fuel industry content with its giant tax breaks. The more competitive they become in the market, the more they rely on friendly lawmakers—who often rely on donations from the fossil fuel industry—to further tilt the playing field in their favor. There's no better example of this than a series of bills that came up last year in the Texas Senate. For more than a century, Texas has handled electricity generation and distribution differently from other states. By remaining independent from the rest of the US grid, it's avoided federal regulations. For a long time, Texan politicians held up their deregulated grid as an example of the state's free-market, Don't-Tread-on-Me values. In 2019, Ted Cruz even tweeted, "Success of TX energy is no accident: it was built over many years on principles of free enterprise & low regulation." His goal, he added, was to "export this recipe for success" to other states.

Thank goodness that never happened. Just two years later, in February 2021, Texas's power grid failed spectacularly. Brownouts and blackouts during the middle of a winter storm left 4.5 million homes without power, caused at least fifty-seven deaths, and cost individuals and businesses more than $195 billion in property damage. Senator Cruz famously fled his state for a Cancun vacation, but he wasn't the only Texas politician with some explaining to do. An investigation conducted by the *Texas Tribune* and ProPublica found that "the state's regulators and lawmakers had known about the grid's vulnerabilities for years but repeatedly prioritized the interests of large electricity producers rather than force expensive changes."

In other words, rule makers in Texas had gamed the system so that it benefited a few large companies rather than tens of millions of Texans. Remarkably, in the years that followed, they doubled down on that strategy. Only this time, the beneficiary was the natural gas industry. Many gas plants failed during the 2021 grid failure, yet under the guise of ensuring reliability, in 2023, the Texas legislature approved the most audacious giveaway to the natural gas industry in history. Their bill would offer companies twenty-year loans with 0 percent interest (basically free taxpayer dollars) to build new gas-fired power plants. If you already owned a gas-fired power plant and agreed not to close it, the state would throw money at you for that, too.

Texas lawmakers claimed that new natural-gas plants would be an "insurance policy," only to be turned on if the grid was under strain. But the idea that the state would spend billions of dollars building gleaming new power plants that they would never use didn't pass the smell test. Outside the Senate, people saw the bill for what it was, an attempt to use taxpayer dollars and the power of the state to prop up the gas industry and protect it from market competition.

Other proposed bills introduced that session, such as one that would require extra environmental impact reviews and public hearings for renewable projects, but not for oil and gas, didn't pass. But they're clear signs of where the fossil fuel industry is heading. And leading Republicans—including several former high-ranking officials from the Trump Administration who would be on the short list for jobs in a future GOP White House or executive branch—have announced an even more ambitious plan: rewriting the rules to favor oil and gas at a nationwide scale.

A report running over one thousand pages and released by the influential Heritage Foundation details "Project 2025," a roadmap for Republicans to follow if they win the presidential election this year. Their agenda would use the full force of government regulation to attack renewable energy and boost fossil fuels. In the first days of a Republican president's first term, they'd block the expansion of the electrical grid for wind and solar energy; close the Energy Department's renewable energy offices; bar states from deciding to adopt California's greenhouse gas emissions limits; and eliminate the EPA's Office of Environmental Justice and External Civil Rights.

And that's just the tip of the iceberg. While Republican voters are increasingly concerned about climate change, and looking to their elected officials to do something about it, the Republican Party remains fully committed to the fossil fuel industry. It's as though, instead of working together to reduce pollution after the deadly pea-soupers of 1952, one half of British politicians were determined to pass pro-fog bills into law. Whether Project 2025 goes into effect will be decided by the results of elections. But this isn't about politics. If we turn the federal government into a subsidiary of the fossil fuel industry—kneecapping their competition in the market and rewriting the rules to boost oil and gas companies' profits—it's going to hurt all of us, Republicans, Democrats, and Independents alike.

Also, while Donald Trump may be the king of climate conspiracy theories, it would be a mistake to blame an entire pro–fossil fuel, anti-market agenda on just one person. When it comes to climate, Trump is a symptom as much of a cause of the madness that has affected the elites in the Republican Party. You can't claim to be in favor of free markets if you try to strangle an industry just because it threatens your campaign donors. You can't claim to be "America First" while trying to keep cleaner, cheaper, safer energy out of Americans' hands. And you can't claim to be tough enough to stand up to countries like China while surrendering our competitive advantage in one of the world's fastest-growing industries.

In a strange way, it's encouraging that the country's most powerful conservative think tank—not to mention the more than 400 conservative elites who helped draft Project 2025—are willing to abandon even the pretense of small-government principle to attack clean energy. There's no clearer sign that oil and gas can't win in the market. But if a Republican wins the White House in the upcoming election, it's hard to overstate the harm it could cause to people both in and outside the United States. Even if Project 2025 can't turn back the clock on the clean energy revolution—and it probably can't; that train has already left the station—conservatives' plan to go to regulatory war for oil and gas could delay the clean energy revolution by years. And we don't have years to waste.

It is within the legal rights of fossil fuel executives to try to game the system, using taxpayer dollars and the regulatory power of the government to enrich themselves and give their companies an upper hand. But the rest of us don't have to sit idly by. We can work backward. First, we can figure out what kind of standards we'll need to meet if we're going to prevent the worst impacts of climate change. Then, we can create new and better rules that reflect those standards, unleashing the power of American business for good as we work together to protect our planet.

Setting the rules will not be easy. Figuring out which policies unlock innovation and competition is a complicated task, with different interests always jockeying for position. But in a democracy, each of us has the power to help decide who writes the rules.

The fossil fuel industry is growing desperate—and fighting with everything they've got. But with enough climate people on our side, we can win this fight. Just as we beat the pesticide companies, just as we made cars safer and more efficient, we can create a cycle of innovation that delivers on the promise of the cleantech revolution. And we can do it far more quickly than most of us today realize.

CLIMATE PEOPLE

Mary Nichols

O ver the past fifty years, climate people have done important work at all levels of government. But few government officials have done more to protect the planet than Mary Nichols.

Mary's first job out of college was at the *Wall Street Journal*, where she was one of the first female journalists the paper hired. But Mary soon realized she didn't want to report on the issues of the day—she wanted to have a more direct impact. She left journalism and went to law school. After graduating, she moved to Southern California, where she began working for the Center for Law in the Public Interest.

At that time, smog was bad throughout the area. But it was especially bad in the city of Riverside, and in 1972 local leaders asked Mary's law firm to represent them in a lawsuit against Los Angeles for failing to limit it. Mary suggested going even bigger: suing the federal government for not upholding the Clean Air Act, which had been signed into law just two years earlier. The city agreed. So, only months after passing the bar exam, twenty-seven-year-old Mary sued the EPA—and won.

The case pushed Mary into the environmental movement, and a few years later she was asked by Governor Jerry Brown to join the board of the California Air Resources Board, or CARB, an agency signed into law by then-governor Ronald Reagan in 1967 to combat air pollution in the state. She served on the board for four years before becoming chair in 1979, helping lead initiatives like removing lead from gasoline and limiting nitrogen oxide pollution from tailpipes.

After leaving CARB, Mary went on to serve as the top clean air official at the EPA during the Clinton Administration, and later as California's secretary for natural resources. In 2007, Governor Arnold Schwarzenegger asked her to return to her old position as chair of CARB, and she agreed, going on to do so for thirteen years.

Under her direction, the air resources board used a cap-and-trade program to implement California's landmark 2006 climate legislation, cutting greenhouse gas pollution by 14 percent and creating a model for a clean-energy transition that states and countries have followed. She expanded California's clean automobile rules and helped the Obama Administration develop national fuel economy standards based on them. Later, when the Trump Administration moved to revoke these rules, she got five major automobile manufacturers to commit to sticking to California's high standards anyway.

Throughout her tenure as CARB chair, Mary also helped oversee and update California's Scoping Plan, the state's plan to combat climate change. The most recent update, in 2022, will make the state carbon neutral by 2045, reducing air pollution by 71 percent while creating four million new jobs in the process, and once again offering a blueprint for the rest of the world to effectively act on climate.

Mary has been praised by regulators who worked under her, governors of both parties who appointed her, and even officials from car companies who battled with her. But it is former EPA administrator Lisa Jackson who best summed up her work:

"If, as Supreme Court Justice Louis Brandeis claimed, states are the laboratories of democracy," Jackson said, "then Mary Nichols is the Thomas Edison of environmentalism."

CHAPTER XI

MEASUREMENTS MAKE MIRACLES HAPPEN

t was Inauguration Day, 1961. In the back of a black Lincoln limousine were two American presidents, one ending his second term, the other just beginning his first. It was also the first time one veteran of World War II handed off the presidency to another. The outgoing commander in chief, Dwight Eisenhower, had served as the Supreme Commander of the Allied Expeditionary Force in World War II. The other, John F. Kennedy, had been a junior officer in the Pacific, at one point given up for dead after a Japanese destroyer had sliced his boat in half.

The two men had a frosty relationship. But now, the younger man asked a question, not as a politician but as one veteran to another. What, he wanted to know, had given the Allies the edge over the German forces on D-Day?

Perhaps JFK expected the great general to mention an intelligence triumph, or brilliant strategic planning, or a newly developed piece of military equipment. Instead, Ike looked at the former lieutenant sitting next to him.

"We had better meteorologists."

Measurement is rarely the most exciting part of solving a tough problem, or of winning an epic fight. But it's often the most important. In the previous two chapters, I've talked about two different components of climate capitalism: better technology and businesses, and better policies and rules. As crucial as all these things are, though, they all become nearly impossible without better ways to measure what is happening in our world, and to our climate. It may not be the sexiest element of business, but if we're going to make the switch from oil and gas to clean energy, there's a good chance that accounting will be our secret weapon.

It's a business cliché, though a true one, that "You can't manage what you can't measure." When it comes to climate, I like to think about the more optimistic version of that idea. If you can figure out ways to measure the seemingly-impossible-to-measure, you can achieve seemingly impossible goals.

That's what Eisenhower was talking about in the car with JFK. At the start of World War II, predicting how powerful the waves would be on a given beach at a given day and time was considered practically impossible. From the start of the war effort, the Allies knew that to win they would eventually need to conduct amphibious landings—not just in France but in Italy and throughout the Pacific. But the best they could do, at the beginning of the war, was order an assault and hope for flat seas. With armies relying on bulky landing craft, and a large number of American troops unable to swim, this inability to predict the size of ocean swells seemed certain to hobble Allied operations and doom countless young men. It was not just inconvenient but deadly.

Two scientists, geophysicist Walter Munk and meteorologist Harald Sverdrup, refused to accept the conventional wisdom that ocean swells were unmeasurable. Together, they led the Allied effort to figure out how to measure waves days or even weeks in advance. For nearly two years, Munk, Sverdrup, and their team of researchers and scientists gathered weather data—sometimes from government stations, sometimes from private sources such as airlines—and matched it against ocean conditions weeks later and across thousands of miles. They tracked every possible factor: wind speeds, storm duration, the size of hurricanes and typhoons. Finally, in 1943, they developed a formula to predict ocean conditions in advance.

It worked. By the time Eisenhower and the other D-Day planners considered when to launch the Allied invasion, they weren't just armed with ships, tanks, and personnel. They had measurements on their side. Initially, the landing was scheduled for June 5, but after consulting with his meteorologists, Eisenhower postponed the invasion to June 6, when forecasters predicted the sea would be relatively calm. Had the Allies attacked on the day originally chosen for the invasion, it's hard to know just how many lives the rough seas would have claimed. Instead, Eisenhower and his forces were able to get 133,000 men ashore in twenty-four hours. For Nazi Germany, it was the beginning of the end.

Today, the fight of our lifetimes is the fight against climate change. Just as Eisenhower relied on meteorologists to give him the edge, and to help him make the right decisions when the stakes were at their highest, better metrics and information can help us stabilize the planet before it's too late.

The most important piece of information we can measure when it comes to climate is pollution. Today, thanks to the work of brilliant and dedicated scientists, we can track how much carbon fossil fuels have been released into our atmosphere on a global level. We can measure annual emissions, emissions by country, or emissions by year. That's important. It gives us a big-picture sense of what oil and gas are doing to the planet, and it allows us to know whether, as a society, we're meeting our pollution-based goals.

When you zoom in, however, the picture gets murky. In certain very specific cases, such as a plane ride, we can measure emissions at what

I think of as a pinpoint level. We can tell how much pollution each flight, or even your seat on a flight, releases into the atmosphere. But for the most part, measuring pinpoint emissions today is considered as far-fetched and impossible as predicting wave heights in 1939.

For example, while I can tell you with reasonable confidence the approximate combined annual emissions from the American agricultural industry, if you asked me, "Which farms and fields, specifically, emit the most carbon?" I wouldn't know the answer. Right now, no one would. Similarly, as a consumer, you don't know much about the climate impacts of most products you purchase. Let's say you buy a pair of jeans. A huge number of factors determines how much carbon they're responsible for emitting. How is the cotton harvested? What kind of manufacturing process is used to make them? What kind of dye gave them their color? Is the sewing machine used to stitch them powered by coal, gas, or renewables? How are they shipped to your house, and from where?

You might think that answering these questions in detail just isn't possible. But we're a lot closer to a world of zoomed-in climate measurements than most people realize. And once we can accurately measure pinpoint emissions, we'll be able to accomplish some very big things.

First, we'll be able to give individuals, businesses, and countries the power to make better choices. Earlier, I wrote about how obsessing over your victory garden isn't a useful strategy, and I stand by that. But at the same time, more and more consumers want to buy things that are helping rather than hurting the planet. Imagine if when you buy a hamburger, or a shirt, or a piece of jewelry, you could know the carbon cost not just of that industry but of that specific item. It would make climate-conscious shopping so much easier.

Which means it would make climate-conscious business easier, too. Right now, with a few exceptions, it's hard to get credit among consumers for producing less harmful stuff. But if you could say, "The jeans I sell have a 20 percent lower carbon footprint than the competition's," you could gain a big advantage with today's consumers. You could sell them things they really want.

The second big benefit of better measurements is that they allow you to accurately price your products. It's often said that in the American system, and indeed, in most capitalist countries, we privatize the gains and socialize the losses. While that's frequently true, when it comes to climate, I'd put it another way. Right now, polluters get a huge discount. That's true not just for companies, but for individuals. I have a friend who regularly flies his private jet from New York City to California and back just to play a single round of golf. That spews roughly 22 metric tons of carbon into the air, more than five times the typical person's average *yearly* emissions, just for eighteen holes.

All that carbon pollution isn't just bad for the planet—it's expensive: climate change requires us to pay for everything from cleanup after natural disasters to infrastructure that needs to be built or upgraded as sea levels rise to lost wages from hourly workers forced to stay home during heat waves. But my friend won't pay those costs, at least not directly. Governments will. Which means taxpayers will. Which means we all will. Right now, my friend isn't just flying a private jet across the country to play a round of golf. He's flying a private jet across the country to play a round of golf *at a huge discount*. He pollutes, and we pick up the tab. I probably can't convince my friend not to fly on a private jet. But if he wants to fly private, he should start paying the full cost of his flight. If the prospect of paying that more accurate cost would make him decide to fly commercial instead, that's fine too.

The problem, however, is that for most goods and services, it's not yet possible to figure out exactly what the real cost is. With my friend's flights, we can get a pretty good idea. But what's the climate cost of collecting a fleet of antique cars, or buying a new yacht, or throwing a big party at your house in the Hamptons? I'm not against having fun. I'm obviously not against having money. And if you have a lot of money and want to buy something luxurious, that's part of how our market-based system works. But if you're buying luxuries, the rest of us shouldn't have to help you pay for them. The only way we can make that happen is if we can better measure pinpoint climate costs.

The fossil fuel industry will, with a straight face, argue that requiring my friend to pay a fair price for his private-jet golf trip is elitist. They'll also claim that if we stop socializing climate costs, everyone will pay more for everything. But that's not true. It's the opposite. According to one study of Americans' carbon emissions, each year, the wealthiest 10 percent are responsible for more than five times as much carbon pollution as those in the middle. On average, an American from a household making $4 million per year is responsible for more carbon pollution in fifteen days than a person making $10,000 or less—in other words, from the lowest-earning 10 percent—emits in their entire lifetime. Yet the effects of that pollution hit middle- and lower-income people the hardest.

In essence, socializing climate losses is a giant, secret tax break for the country's richest people. As one of those people, trust me, we don't need another tax break. We're doing fine.

The good news is that once we can better measure climate costs, we can flip the current system on its head. If you can fly on a private jet to play a round of golf, you can afford not just to pay for your own costs, but to help pay for a restaurant hostess or teacher who has no choice but to drive to work. Once we can more accurately measure carbon pollution, it's not hard to imagine a future where working-class Americans get a huge tax break to reflect the fact that they're not the ones who are really driving climate change.

The oil and gas companies, on the other hand, *are* driving climate change—and another benefit of more accurate measurement is that it will allow us to hold them accountable for their actions. Today, we're in almost the exact same situation that we faced with cigarette companies a few decades ago. We know those companies lied about the harm their products were causing. We know they fought to keep us addicted to their products. And the costs of a lung-cancer epidemic fell on patients, and on taxpayers through Medicare and Medicaid. But eventually, better measurements helped hold the tobacco companies accountable. In the late 1980s, researchers interviewed 35,000 people in an attempt to figure out tobacco's cost to public health programs. By compiling data on how much people smoked, smokers' typical age, medical expenses for smoking-related diseases like emphysema or heart attacks, and smokers' typical health coverage, they were able to create a national model for smoking-related public health costs. Then the researchers dug even deeper, applying their model to individual states. The result was shocking, even to those who knew cigarettes were dangerous. Smoking-related expenses made up between 8.6 percent and 19.2 percent of each state's public health spending. In part because states were able to quantify the costs associated with cigarette companies' decades of lies, they were able to do what a few decades earlier had seemed impossible—beat Big Tobacco in court.

The same thing will happen with fossil fuels. In fact, that's one reason I wouldn't want to invest in oil and gas companies right now. They're massive liability risks. Over the past five years, more than a dozen states and municipalities across the country have sued oil companies, arguing that they knowingly deceived the public about climate change and its effects. Different courts and juries will rule in different ways, and it's likely that, given its massive political clout, the fossil fuel industry will win some cases. But since climate change affects every state and municipality, climate people don't have to win every lawsuit to start holding oil and gas accountable. I'm not a lawyer, but it seems to me that the fossil fuel industry, for all its arrogance and smugness, may soon reap what it's sowed. ExxonMobil, for example, was projecting global warming increases with an accuracy within one degree Fahrenheit as far back as the 1970s. They knew exactly what they were doing. It's hard to argue that they're not liable for their actions given their decades of lies.

The all-important question will be the extent of the damages. And that's also where better measurements come in. If we can say, "Oil Company X produced 5 percent of cumulative greenhouse emissions, and climate change has already cost this state $Y million, and will cost another $Z million over the next several decades," we'll be able to arrive at a starting point for how much that company ought to owe. And if the lawsuits against Big Tobacco are any guide, real damages are likely to be even higher, since they'll include additional punitive penalties for deceiving the public.

The better we can measure the sources of emissions, and the precise costs imposed by those emissions, the sooner we can hold the fossil fuel industry accountable—and shift the burden for cleaning up the mess they've caused from innocent taxpayers back to the companies that knew they were hurting our country. So it should surprise no one that fossil fuel companies want to make better measurement of climate emissions as difficult as possible for as long as possible.

Earlier, I discussed how methane leaks can cause so-called clean natural gas to become one of the dirtiest forms of energy. If you lose just 3.2 percent of your natural gas to leaks, it becomes the dirtiest fossil fuel in the energy business, even worse than coal. Just over a year ago, I met with some natural gas executives in Houston, to try to understand how they see the world. At one point, I asked them, "Why don't you invest in more technology that will help you identify natural gas leaks? You'd save money, and you'd be able to demonstrate that your energy is actually clean. It's a win-win."

The executives talked in circles for a while. That's when it dawned on me. *They didn't want to find out.* If they can't identify their leaks, they have plausible deniability. They can still claim that natural gas is cleaner than coal and point to theoretical numbers in the lab, without having to worry that (as is likely the case) the real-world data points tell a very different story. Even better, from their perspective, is that no one can make them pay for their mess, because no one knows exactly how big a mess they've made. In other words, they're willing to lose 3 percent or more of their product in order to plead ignorance about the harm they're causing. That's how valuable plausible deniability is to them.

The same dynamic occurs on an even larger scale with what's called "Scope 3 Emissions." Scope 1 emissions are the byproducts of any fossil fuels you burn directly. Let's say I run a dog-food factory. Any gas I burn to cook or dry my product, or seal it into cans, or ship it to customers using trucks I own, falls into Scope 1. Scope 2 emissions is the indirect carbon pollution that results from generating electricity from my business—everything from lighting to air conditioning to keeping the servers running. Both these types of pollution are, relatively speaking, measurable. Usually, when people and companies pledge to cut their emissions to zero within a certain timeframe, they're referring only to these two categories.

But Scope 3 emissions are a much bigger piece of the picture—because this category includes the full supply chain. In my hypothetical dog-food factory, we'd be talking about ingredients for the food itself; aluminum for the cans; paper for the labels; desks and tables and chairs for the office; any shipping done by FedEx or UPS, as opposed to by my own company fleet. It's impossible to know how many tons of carbon pollution my factory is responsible for if we only measure the impact of my production line. We need to measure the impact of my supply chain, too. In fact, the Carbon Disclosure Project, an international nonprofit that calculates environmental impacts for nearly 20,000 organizations, estimated that Scope 3 accounted for an average of 75 percent of the average company's carbon pollution. But today, even in companies that pledge to bring their emissions down dramatically, Scope 3 isn't measured.

A lot of people in business will tell you that's because measuring Scope 3 emissions is impossible. But what they mean is that they don't want to do it. It's not hard to understand why. Today, there's a big loophole for companies that want to be seen as climate-conscious but don't want to make major changes: they outsource pollution to other companies and suppliers, call it Scope 3, declare it unmeasurable, and brag about how low their Scope 1 and 2 emissions are. Even worse, ignoring Scope 3 emissions gives companies incentives to ship production to businesses that cut corners with dirty energy. China, for example, burns more coal than any other country in the world, yet companies whose supply chains rely on Chinese factories don't have to acknowledge the existence, let alone the costs, of all that pollution.

If we could measure Scope 3 emissions, businesses would be forced to put pressure on their supply chains—including, crucially, suppliers in countries where governments are less likely to measure and regulate carbon pollution. This would help make sure that companies who cut corners don't get an unfair advantage over those that play by the rules. It would require countries like China to reconsider whether coal is really as cheap as they think it is, and to accelerate their development of renewable energy in order to stay competitive in the global market.

And that's a good thing. Companies should meet their emissions goals by cutting emissions, not by exploiting loopholes. At the same time, taking Scope 3 emissions into account would benefit suppliers, and even entire nations, who want to do things the right way but get no credit for it. Measuring Scopes 1 and 2 is a lot better than nothing, but we need to go further if we're going to hold bad actors accountable, reward those who do the right thing for the planet, and accurately reflect the costs of carbon pollution in the decisions we make.

Today, some of the innovations and companies I'm most excited about, and which I've invested in through Galvanize, aren't directly reducing pollution by sucking carbon out of the air or by generating cleaner electricity. Instead, they're indirectly reducing pollution by making the unmeasurable measurable.

A great example is a company called Regrow. You probably don't realize it, since the American food system tends to use wheat, soy, and corn as staple grains and starches, but rice farming has an outsized impact on global emissions. Studies have found that, per ton, emissions from growing rice are twice those of corn, and nearly five times those of wheat. That's because unlike other crops, most rice is grown on flooded land. Deprived of oxygen, the soil in a rice paddy is a perfect environment for naturally occurring microbes to grow, and these microbes release huge amounts of methane as waste. Some estimates say that as much as 1.8 percent of all greenhouse gas emissions come from rice.

But knowing how much methane is emitted by a single rice paddy has generally been considered impossible. This, in turn, has meant there's been no way to reward farmers who change their practices to reduce pollution—which means there's less of an incentive for them to do so.

Regrow is working to change that. Using a model that combines satellite imagery with on-the-ground verification systems, they allow researchers to test different types of methane-reducing strategies and quantify their effects. This, in turn, allows farmers to quantify how much pollution they're responsible for, based on how much land they're farming and which methods they're using. That's important because once farmers know how much greenhouse gas they're emitting, they can adopt new practices that reduce their emissions, which in turn will make them eligible for credits or payments.

Another company Galvanize has invested in that's using measurements to transform agriculture is called Arable. Earlier, I mentioned Gabe Brown, the North Dakota farmer who uses regenerative agriculture to eliminate fossil fuel–based fertilizers and increase the amount of carbon sequestered by his soil. Unfortunately, the cost of soil testing remains high, and measuring annual increments in soil carbon is extremely challenging. For most farmers—especially those farming at a large scale—it doesn't make economic sense to measure carbon content or carbon pollution by going from field to field and taking samples. But Arable has invented a sensor that can be placed in a field and left there. Data from the sensor give farmers access to information about their environment and crops they wouldn't otherwise have, allowing them to manage their resources in ways that cut pollution and lower costs while maintaining, or even growing, their yields.

The beauty of companies like Regrow and Arable is that they're working to make things simple and inexpensive. Their goal is to build a future in which finding out how much carbon the soil on your farm contains, or how much methane you're releasing into the atmosphere, is no longer a mystery that takes a lot of time and money to solve. Anyone will be able to do it. Not only that, those who do can save money.

Which brings us to another benefit of better measurements: When we break down big problems into small ones, they become a lot easier to solve.

We've all had experiences in which something goes wrong, whether personally or at work, that feels overwhelming. But when we take a step back and make a list of all the issues that need to be addressed, we can deal with one, then another, then another. Before we know it, the problem isn't so big anymore.

In software development, there's a useful term for this kind of thinking: a "punch list." Let's say you're about to ship a product, but it's just not working. If you say, "This is broken, how do we fix it?" you're doomed. The task is too daunting. But if you make a list of all the little things you need to fix, and you pursue them one by one, it's suddenly easy to get started.

Right now, we can measure enough about the climate to feel overwhelmed. We know the amount of greenhouse gas that burning fossil fuels sends into the atmosphere. We know that as a planet, our emissions are still going up, when they need to be coming down. We know how rapidly weather patterns are changing, how abruptly temperatures are rising, how quickly sea ice and glaciers are melting. We know just enough, in other words, to feel overwhelmed.

That's exactly how the fossil fuel companies like it. They no longer try to pretend that burning oil and gas is good for the world. Instead, they say, "Sure, this will threaten humanity, but having the entire world quit fossil fuel for something cleaner is so hard to imagine that we have no choice but to let the planet burn."

The better we measure, though, the more we understand that we really do have a choice. None of us can "solve climate change." But each of us can solve a small part of climate change. Our job, as human beings, and in particular as Americans, is to answer the question, "Which part of climate change can I solve?" For many people, the answer will involve helping to create measurements. No matter what field you're in, if you deal with budgets, projections, or data of any kind, you already know how essential metrics are to solving big problems. And you probably already understand how to gather and use information in the most effective way. With a clean-energy revolution unfolding before our eyes, there's almost certainly a way for you to take that set of skills and apply it to stabilizing the planet. Nearly sixty-five years ago, John F. Kennedy asked Dwight Eisenhower what gave us the edge on D-Day. As improbable as it may seem right now, I believe today's children will one day be able to ask a version of the same question, from that same place of victory. They will want to know what gave us the edge in the worldwide fight against climate change.

And I think there's a good chance that we answer, "We had better measurements."

CLIMATE PEOPLE

Riley Duren

One of the biggest keys to stabilizing our planet is tracking carbon pollution. While a quick Google search can tell you where it's raining at this very moment around the world, we have no idea where the largest amounts of greenhouse gas are being released.

Carbon Mapper is going to change that.

A coalition made up of NASA's Jet Propulsion Laboratory, the State of California, the University of Arizona, Arizona State University, the Rocky Mountain Institute, and the satellite-imaging company Planet Labs BPC, Carbon Mapper uses satellites and planes equipped with imaging spectrometers for remote sensing to detect, track, and publicize data regarding greenhouse gas pollution. Its goal, says co-founder and CEO Riley Duren, is to be "like the weather service for methane and CO_2."

Riley helped launch Carbon Mapper in 2020. During a thirty-five-year career at NASA, he worked on space shuttle missions at the Kennedy Space Center, served as chief engineer for the Kepler mission that scoured the Milky Way looking for Earth-like planets, and ultimately helped lead the Jet Propulsion Laboratory's work on climate change. This experience led him to Carbon Mapper.

The organization is a testament to the power and importance of collaboration on climate. Technology developed by NASA takes invisible infrared light and breaks it into different colors to identify sources of pollution. Airplanes and satellites outfitted with that technology help gather the data through regional surveys. Scientists at Carbon Mapper process this data and publish it. And the organization works with facility operators and agencies like the California Air Resources Board that take action to address the leaks and nonprofits like the Rocky Mountain Institute that help build support for policy solutions.

Over the past few years, Carbon Mapper has tracked and published data on more than 12,000 greenhouse gas leaks and plumes coming from oil and gas wells, refineries, pipelines, mines, landfills, wastewater treatment plants, and agricultural sites in thirty-three states, and helped stop many of them. In California, for example, Carbon Mapper data led forty-four separate facilities to take action on leaks, keeping as much as 100,000 metric tons of methane, equivalent to 2.5 million metric tons of CO_2, from the atmosphere, while data gathered in Pennsylvania resulted in twelve facilities working to limit leaks, cumulatively decreasing methane pollution by 10 percent.

In early 2023, Carbon Mapper extended its work to South America, flying planes across the region to identify methane emissions from landfills and oil and gas refineries in Chile, Ecuador, and Colombia. Going forward, the organization plans to use satellites to identify leaks across the world. The first of these satellites will track select pollution from a small group of "super emitter" sites like power plants and refineries. More satellites will be able to observe dozens of sites daily, and if all goes well, Carbon Mapper will eventually deploy enough satellites to monitor emissions across the entire planet and quickly release this data.

Gathering this information won't bring down greenhouse gas pollution on its own. "Action needs to follow," says Riley. But with increasing momentum to act on climate, he and the team at Carbon Mapper believe that with better knowledge, transparency, and data, more public pressure and stronger government responses are sure to follow.

CHAPTER XII

BEING RIGHT ISN'T EVERYTHING

So far, we've discussed winning in the marketplace in a literal sense—creating products and services and then building a business that delivers them to consumers. But there's another, figurative marketplace that is often just as important: the marketplace of ideas.

In my experience, many people don't realize the extent to which these two marketplaces influence one another. That's especially true in the climate movement. Sometimes, we think that because we're objectively in the right—factually right because the science supports what we're saying, and morally right because we're fighting to protect billions of human beings from needless suffering—the arguments in favor of our ideas should make themselves.

But the world isn't fair, and being right is not enough—especially when it comes to markets. With nearly a half century of experience in and around the private sector, I've learned that you can't just have great ideas, great technology, or even a great business built around that technology. You have to make your case in a compelling, persuasive way.

Many individuals who are working hard on climate understand the importance of messaging. But as a movement, we consistently fall short. Here's just one example, which is both really small and really big at the same time:

When it comes to measuring the warming of our planet, everyone uses Celsius. The goal of the Paris Climate Accords was to keep warming below 1.5 degrees Celsius. When researchers try to determine what the consequences would be if, as seems almost certain, we miss that target, they say something like, "If warming reaches X number of degrees Celsius, Y will happen." I'm as guilty of this as anybody. If people ask what our current climate trajectory looks like, I'll reply, "I think if we don't act soon, we're going to experience at least three degrees of warming"—I don't even specify Celsius. In the climate movement, Celsius is assumed.

There's just one small problem. No one in America uses Celsius! Think about how bizarre this is. The United States is the richest, most geopolitically powerful country on Earth. With the possible exception of China, it's the country where finding the political will to take significant action on climate is both most important and, thanks to Donald Trump's Republican Party, most difficult. Yet when it comes to conveying the most fundamental information about climate change to the public, we use a language that, for all intents and purposes, Americans don't speak.

Celsius became the standard when discussing climate for a good reason. Researchers and policymakers in every country need to refer to the same measurements, and most countries use Celsius. But those figures need to be translated to allow non-academics and non-policymakers to understand them. Imagine if America's jobs reports were only issued in German, or if the president delivered the State of the Union in French.

But more than being confusing, sticking with Celsius when we talk about climate change comes across as arrogant—playing right into the stereotype of the environmental movement that the fossil fuel industry spends so much time and money to perpetuate. Without meaning to, climate people wind up signaling that a bunch of elites have decided that they understand the world, and you don't, and so they're going to talk about it in a language you don't understand.

I'm not suggesting that climate change will magically go away if Americans switch to talking about global warming in degrees Fahrenheit. I am, however, saying that it's very hard to get people to listen if you insist on speaking to them in a language they don't use.

Implying that climate change is solely the concern of scientists and overeducated eggheads isn't just terrible salesmanship, it's false. I can't imagine any movement more populist than the climate movement. Its goal is to make life better, cheaper, and safer for everyone—and working-class people, who will be hit hardest by the costs of disasters like heat waves and hurricanes, are going to benefit far more than billionaires. Yet the language we use puts our worst foot forward and plays right into the fossil fuel industry's hands.

The oil and gas industry understands exactly how important the marketplace of ideas is to their business. Consider what happened to the once bipartisan idea known as "cap and trade."

The idea behind cap and trade is simple. Companies buy permits that allow them to emit a certain amount of greenhouse gas pollution. They can use those rights—or they can trade them on an open market. If you cut your emissions, you can make extra money by finding a buyer for your unneeded permits. If you can't figure out a way to run your business within your initial greenhouse-gas limits, you can buy extra permits to cover your needs. The idea was that by gradually bringing down the total amount of pollution allowed in the market, we could cut society-wide emissions in a market-based way.

Cap and trade isn't some wild theory. In the 1980s and 1990s, acid rain caused by sulfur dioxide threatened ecosystems across the country and the world. We set up a cap-and-trade system for sulfur dioxide, gradually phasing out how much pollution businesses could emit. Now you never hear about acid rain anymore. We harnessed the power of capitalism to beat it.

So it shouldn't be surprising that in the early aughts, cap and trade for carbon emissions was a popular idea with Democrats and Republicans alike. Former GOP House Speaker Newt Gingrich publicly supported it. President Obama pushed forward a cap-and-trade bill precisely because it was a middle-of-the-road, historically bipartisan policy solution.

But while cap and trade would have made life better for almost everyone, there was one group of people who stood to lose big-time: oil and gas. Lowering the cap on emissions would mean, over time, burning fewer fossil fuels, which means less demand for oil and gas. The only companies that would be hit harder than those that drilled fossil fuels were those that built infrastructure, like pipelines, for transporting them. The owners of one of those companies—Koch Industries—went to war against cap and trade.

Part of the Koch Brothers' strategy involved lobbying. They donated millions of dollars, set up a giant corporate lobbying arm headquartered two blocks from the White House, and funded so-called "grassroots" protests. But they also engaged in the marketplace of ideas. They and their allies began to talk about climate permits not as a market-based program, which conservatives liked, but as a tax, which conservatives hated. They got the entire Republican leadership to pledge not to support any cap-and-trade bill unless all the revenues were offset by corresponding tax cuts. Then they renamed cap and trade "cap and tax." Suddenly, Obama's bill went from sounding like a free-market program (trading) to sounding like a nanny-state program (taxing).

"We turned it into 'cap and tax,' and we turned that into an epithet," said Myron Ebell of the Competitive Enterprise Institute, a conservative think tank. "We also did a good job of showing that a bunch of big companies—Goldman Sachs, the oil companies, the big utilities—would get windfall profits because they'd been given free ration coupons." Think about that. Koch and their allies knew people don't like the big oil companies. So they framed the biggest climate bill in history (at the time), something that would have sent us rocketing toward lower pollution and cleaner energy, as a giveaway to big oil.

I don't admire the ethics of that strategy, but I'm struck by its effectiveness. The Koch Brothers understood that winning in the marketplace of ideas translates directly into winning in the marketplace of goods and services. As climate people, we have so many competitive advantages. We're on the right side of history. We're on track to produce better, cheaper, cleaner goods. We're doing well on a playing field tilted against clean energy, and we will do even better if the field is leveled.

But all those things aren't enough. We need to compete much more aggressively in the marketplace of ideas. Which means that if you're already involved in the marketplace of ideas—if you work in communications, or advertising, or marketing, or public relations—then the climate movement needs your help more than ever.

I'm an investor, not a communicator. I won't pretend that I know exactly which messages will be most effective when it comes to climate. But I do want to step back and suggest a few broad ways climate people can beat the fossil fuel companies, not just when it comes to performance and price, but when it comes to persuasiveness as well.

First, we have to focus on people instead of nature. I can't tell you how many ads I've seen that talk about how melting sea ice is threatening polar bears' access to food. More generally, think about how often the case for climate action hinges on protecting a special landscape, or an iconic animal species, or an island halfway around the world.

Don't get me wrong: I love nature. In fact, I think of climate change as just a broader symptom of humankind's disregard for the natural world. I suspect that what sparked many people in the climate movement to get involved in the fight against climate change was something like my trip to Alaska with my family—a visceral understanding of the way global warming threatens the most beautiful places and species on earth. I also know that nature lovers make up the bulk of donors to environmental organizations.

The problem with the save-the-polar-bears messaging, though, is that it preaches to the choir. If you're trying to raise money from people who already support environmental causes, polar bear messaging might be effective. But if we're trying to build a bottom-up, grassroots movement, we have to recognize that for most people—even those who think polar bears are beautiful creatures—saving the polar bears is not a top priority. Rightly or wrongly, hundreds of millions of human beings are not going to change their behavior so that polar bears can have more ice to fish from.

If we focus on nature instead of people, it also means we're missing opportunities to help people understand just how much climate change is a story of human suffering. From the Marshall Islands to Miami, from the Democratic Republic of the Congo to Dallas, carbon pollution is already taking a terrible toll on our quality of life.

Even the fossil fuel industry and their allies concede this point. After last year's unprecedented heat waves in Texas, the *Wall Street Journal* ran an article whose point was, "It's not too bad, just stay inside from 7:00 a.m. until 10:00 p.m." Popular conservative pundit Ben Shapiro made a similar point. "It's hot outside," he told his listeners. "You know what I can do about that? Zero things. Thank God we have this thing called air conditioning."

Let's consider for a moment that both the *Wall Street Journal* and Ben Shapiro, the same people who told us a temporary COVID stay-at-home order was the greatest assault on liberty in modern history, are totally fine with fossil fuels forcing us to stay indoors all summer, every summer, forever. Basically, the fossil fuel people are saying, "Reduce your quality of life in order to keep using outdated, dirty technologies," and the climate people are saying, "Improve your quality of life—switch to something cleaner instead, that, in most cases, is cheaper, too." We should be winning that debate hands down! But when we focus on nature instead of people, we're not even making that argument, much less winning it.

Another way in which climate people fail to make our case is that we rely on journalists to make it for us. I have the utmost respect for journalists. In fact, my mother started her career working for NBC News, at a time when women had to fight tooth and nail to be part of a newsroom. She taught me that journalism is a calling. But while journalists often do an excellent job of covering climate, their job is primarily to *document* climate change, not to protect the planet from it. And because good reporters don't want to be seen as taking sides on a politically controversial issue, they tend to balance the liberal and conservative points of view rather than highlighting an overwhelming consensus among scientists and experts. Both these tendencies mean that fossil fuel companies are incentivized to make climate as politically controversial as possible.

Which is exactly what they've done. The more desperate the oil and gas industry becomes, the more aggressively they try to make climate a polarizing issue in our already polarized country. It's gotten so bad that Chris Gloninger, a TV weatherman in Iowa, was forced to retire after receiving death threats for mentioning climate change in his reporting about natural disasters. He had his facts right. He was doing exactly what a good local journalist should do—inform people about what's going on in their community. But the fossil fuel companies, along with their political and media allies, have made even basic, objective reporting feel like wading into politics and the culture wars.

In large part because the fossil fuel industry has injected so much political controversy into what shouldn't be a political issue, media outlets are often one step behind the facts. They're comfortable reporting the weather. If that weather is extreme—a heat wave, drought, wildfire, or flood—they'll cover it, too. But when it comes to climate—what's causing the extreme weather, who's responsible, what we're likely to see in the future if we don't act, and how we might prevent it—news organizations are generally risk-averse.

Also, while reporters have access to people, they don't always have access to all the information. I can't tell you how many times I've been talking to a journalist about climate, only to discover that, while they can get all sorts of scientists, politicians, and policymakers on the phone, they don't know key details about what the latest research says, or what's happening in the tech and business world. Reporters are on deadline, and climate is often just one of many areas they cover. They can't go to all the meetings and conferences. They're not in the room when early-stage cleantech companies ask for funding.

None of this lets journalists off the hook. They have an incredibly important role to play in our society, and we all rely on them to do their difficult jobs as well as possible. But we can't assume that journalists— even those who work incredibly hard and do their homework—know everything climate scientists know about climate or that they feel free to express what they do know in a way that fully conveys the urgency of action.

Journalists have an important role to play in helping people understand the effects of climate change and the clean-energy revolution taking place. But as climate people, we can't outsource the hard work of messaging to them. There are endless ways we can improve our ability to actively compete in the marketplace of ideas, and, I hope, a near-endless number of climate people who are ready to help think up those improvements. But for now, let's focus on three important areas: naming, simplicity, and what people call "branding."

During my career as investor, I learned that one of the most helpful things you can do in business is to come up with the right name for what you're doing. I learned that lesson from David Swensen, the genius investor who for decades managed Yale University's endowment. When

Fleur Fairman and I originally asked him to invest money in Farallon, he turned us down. "We don't like the way your structure works," he told us frankly. "If you make money, you keep 20 percent of it as a fee, and if you lose money, you two are going to start a new business because you're not going to want to have to make the money back before you get your 20 percent."

I was ready to accept defeat, but Fleur was furious. "That's a bunch of bullshit!" she said. "If that's who you think we are, we don't want your goddamned money anyway." David replied, "Good. You got it." He was the first institutional investor who gave us money. It changed our company, and made it—and me—a lot more successful.

But that wasn't the only way David shaped Farallon. During my first years in business for myself, I thought I was doing essentially the same thing I'd done at Goldman Sachs, point-to-point investing around specific, predictable events. We were expanding our horizons intellectually and doing things we'd never done at Goldman—everything from defaulted debt to real estate to film finance—but I still thought of myself as being in the risk arbitrage business.

But David said, "I think you guys are sort of a new category. And it needs a name." The phrase "absolute returns" came up in conversation, and it stuck. It didn't change anything we were doing. Yet there was a huge difference between saying "we do risk arbitrage but slightly differently" and "we do absolute return investing." Instead of being in the "risk" business, we were in the business of absolute returns.

It's the same with venture capital. People have been lending money to their brother-in-law's crazy idea since time immemorial. But give it a title, and help people understand it, and suddenly you have an industry. That's what happened to us. When we found the right name for our work, suddenly more people wanted to be part of it.

Over the last decade, there's been at least one huge win for climate people in this regard. Twenty years ago, we talked about greenhouse gas almost exclusively as "emissions." "Emissions" is a neutral, scientific word. Yet greenhouse gas is not at all neutral, and its effects aren't matters of scientific theory—they threaten life on earth as we know it. Today, while most people in the climate movement, including me, haven't given up the word "emissions," we've also begun to refer far more often to "carbon pollution," which is just as accurate a way of talking about what oil and gas companies are pumping into the atmosphere. The change in language helps people immediately understand what the stakes are, and why we need to act.

Climate people need more naming wins like this, and the sooner, the better. You may have noticed, for example, that throughout this book I've used "greenhouse gas" instead of "greenhouse gases." That's because the former sounds like what it is—a type of dangerous pollution—while the latter sounds like something you'd learn about in chemistry class. It's the kind of small change in phrasing that I believe can make a big difference.

Another potential win involves names in the most literal sense. Today, we give names to tropical storms and hurricanes, but heat waves—which kill far more people than hurricanes—remain completely anonymous. Recently, a movement has been growing to assign names to heat waves just as we do to storms, to underscore the need for communities to prepare for them and to make clear that what we're talking about is not just a few days of unseasonable warmth but a category of dangerous extreme weather. I think that's a great idea.

We can also do a better job of naming things as we explain the harm that the fossil fuel industry is causing—and its grip on our market economy. Oil and gas loves to act as though they're free-market winners, and climate people often use language (such as the "green premium" I mentioned in the last chapter) that concedes this point. But it's not true—and we should say so. How you precisely measure the full cost of a barrel of oil is complicated enough to be the subject of another chapter, but the short version is that, as I've mentioned, fossil fuel companies get all sorts of big tax breaks that help them drive down the cost of their products and drive up their profits, and then they have the disasters and public-health crises caused by their pollution cleaned up by governments at taxpayer expense. Climate people often call these handouts "subsidies." But there's a better, more accurate name for money an industry receives from a government without which it couldn't keep its business afloat. The fossil fuel industry is the recipient of a massive, perpetual bailout. We shouldn't be afraid to say so.

Climate people could learn a lot about messaging from Warren Buffett. Warren has a gift for explaining his ideas about investing in the simplest possible way, so that anyone can understand them. Some people say smart things. Some people say memorable things. Warren is consistently saying both.

Here's one of my favorite Warren Buffett sayings: "The first rule of investing is 'Don't lose money.' The second rule is 'Don't forget the first rule.'" How perfect is that? What Warren is describing here is actually a really complicated philosophy—one that many people disagree with. He's saying that investing isn't about being a gambler, it's about doing your homework, understanding the value of different industries, know-ing what creates a great business, and then making long-term, low-risk bets based on your accumulated knowledge and the power of compound interest. Somehow, he manages to sum all of that up in fewer than twenty words, in a way that anyone can understand and remember.

Climate people should always be looking not just to share the truth, but to share the truth in the simplest way. That's how we can bring more people over to our side as quickly as possible. Too often, those within the climate movement get excited about complexity. In most cases, their intentions are good—these are curious people who have worked hard to understand an issue with untold moving pieces, and they want to demonstrate both their authority and the depth of their understanding.

But most people's brains turn off when they hear a twenty-point list of variables and statistical ranges. Even worse, our tendency, as listeners, is to take complicated or technical language personally—to view it as condescending rather than just wonky.

But truth is not just about what we say. It's about what we do. Which brings me to the final area of messaging I want to discuss as we think about winning in the marketplace of ideas. Like a lot of people, I roll my eyes at the phrase "brand authenticity." To me it sounds manipulative. More than that, though, I think it's too clever by half. There's no such thing as brand authenticity. There's just authenticity. If you say one thing and do another, no amount of clever branding will convince people that you can be trusted. If, on the other hand, your words are consistent, and your actions are consistent, and your words and actions are consistent with each other, you'll naturally end up with a strong brand whether you consciously think about your branding or not.

Interestingly, it was Warren Buffet (in another example of the power of naming) who helped popularize the idea of "brand equity." Basically, while a company's brand isn't tangible, it's real. It has value. And if a company isn't careful, they can lose the value of their brand. In many ways, that's what's happened to American fossil fuel companies in recent years. For a very long time, they stood for innovation, energy, and exploration; they represented the cutting edge of human-made miracles that led to a brighter future. Now they're viewed as jeopardizing our future, not improving it. Even worse, they are standing against, not for, innovation. Oil and gas is still hugely profitable, but it's got a well-deserved branding problem.

Unfortunately, the climate movement has a branding problem of its own. The stereotype of climate groups—and of environmental groups before them—is that they're made up of wealthy, overeducated white people who don't understand how the real world works. Unfortunately (and I recognize that, business credentials aside, my personal demographics don't exactly help solve this problem) the stereotype of climate people often rings true. We talk about protecting the planet, and about America's ability to lead the way. But when it comes to income, occupation, education level, and race, our movement doesn't accurately reflect the planet we're trying to protect or the country that we believe can lead the world in this fight.

This is just one more reason—among many—that I believe in environmental justice. The fact that poorer countries will bear a disproportionate burden of the impact from climate change is unfair. It's similarly unfair that within this country, marginalized communities suffer the most from oil-and-gas-related pollution, even as their members are often the least able to protect themselves from the impacts of climate change. Addressing these inequities is the right thing to do, in and of itself. But beyond that, the fight for environmental justice aligns the climate movement's actions with its words. We say that we're fighting for everybody— from coal miners in Appalachia to Black farmers in the Mississippi Delta to my own grandchildren. Working toward environmental justice is one of the most important ways we can walk the walk.

There are, of course, disagreements among climate activists, policymakers, scientists, and entrepreneurs. Protecting our planet from climate catastrophe raises a seemingly limitless number of difficult questions, and it's only natural that different people answer those questions in different ways. The problem arises when organizations appear to be putting their own interests, or their donors' interests, ahead of the broader mission climate people share. Movements are never perfect, and the climate movement is no exception. But the more consistent our actions and words are, the stronger our brand will be—and the greater the contrast will be with the hypocrisy and inauthenticity of the fossil fuel companies.

I've seen firsthand that the right messages on climate can have an enormous impact. In the 2016 elections, an organization I started, NextGen Climate Action, registered more than 1 million young people to vote, and a large percentage of those young people turned out at the polls. The reason was simple: we engaged them on climate. A lot of people my age don't think about the fact that for people in their teens and twenties, climate change has been occurring for their entire lives—and it threatens to destroy, or at the very least irreparably damage, their futures. Today's young people are not just receptive to smart, honest arguments from climate people, they're desperate for them.

Back in 1990, even if you were a great communicator on climate, you would have had a very difficult time getting anyone to take climate change seriously. Today, on the other hand, there's never been more opportunity to reach people, to help them understand the stakes and the science, and to convince them to take action. The fossil fuel companies aren't going to stop competing in the marketplace of ideas—in fact, the less competitive their products become and the more it becomes clear how much harm they're doing, the more money they're going to throw at lobbying, advertising, and talking points to keep themselves in business. But over the next few years, as incredible breakthrough technologies and companies drive the cleantech revolution, you won't have to be an inventor or an entrepreneur to do your part. If you work in marketing, or public relations, or media, or communications, or branding—in short, if you're part of the marketplace of ideas—we need your help.

CONCLUSION

My daughter Evi has worked with me on climate issues since 2021. A few years ago, she asked me for some pretty serious advice. "Dad," she said, "I'm considering not having kids. I read all this news, especially on climate, and it just seems unfair to me to bring children into this world. What do you think?"

It's a question, and a fear, I hear a lot these days—especially from young people who work on climate. They're inundated with frightening, sometimes tragic news. Knowing what we know, they wonder, how is it ethical to condemn a child to live on a planet increasingly defined by disaster?

But I told Evi what I tell them. Of course you should have kids! Having a child is the ultimate statement of optimism. It's a way of saying, "I want to keep participating in the future of the world, and I refuse to give up."

I hope that by now you're a climate optimist, too. We have a huge fight ahead of us, and the fossil fuel industry isn't going to fade away quietly. But the tide has already turned. Thanks to new technology, increased public awareness, new laws and rules, and new ways of measuring and understanding the impact of greenhouse gas pollution on our planet, the clean energy revolution hasn't just begun—it's become unstoppable.

Oil and gas companies will fight tooth and nail to delay it. But even they know it can't be reversed.

Still, as much as I would like to say that we don't have to worry about climate change harming, and perhaps even defining, the future of a child born today, that just wouldn't be true. Optimism doesn't mean blissful ignorance, and looking at the science, it's clear that we're far from out of the woods. In fact, we're still making the crisis worse. Each year that we pump more greenhouse gas into the atmosphere than we can sequester, we further disrupt natural systems that human beings have depended on for millennia, unleashing even more extreme weather events—from heat and drought to hurricanes and wildfires—that have already become frighteningly routine. To put it simply, the longer we continue to burn fossil fuels on a massive scale, the worse our future will be.

And not just for as-yet-unborn generations. Recently, I had a conversation with a friend about my age. He said, in essence, that climate change is real, and caused by human beings, but that it's going to be the next generation's problem. All I could think was, *seriously?* Tell that to people in Arizona, Texas, Vermont, and Hawaii. Or Libya, Greece, Pakistan, and Canada. A warming earth claimed victims in each of those places last year alone—and the full list of such places would be far longer. Climate change is here, and even if it doesn't kill you or drive you from your home, it's going to affect you no matter where you live.

At the same time, however, the global nature of the threat also means that there's a worldwide opportunity to help defeat it. Nowhere is the adage that "anyone can make a difference, and everyone should try" truer than in climate.

Which brings me to the other reason I'm 100 percent sure that bringing children into this world is still the right thing to do. Embedded in the idea that the best response to a warming world is to abstain from having kids is the assumption that a human life must, by definition, be a burden on our planet. But that's just not true. Every day, I work with people who will leave the earth far better than they found it. Earth, and the 8 billion human beings who inhabit it, are very lucky that these people exist and that they've chosen to spend their time in the way that they have.

Throughout this book, I've highlighted some of the people who have devoted their lives—often after a mid-career change in focus—to fighting climate change. Their activism, ideas, and businesses are making the clean energy revolution possible. But while their stories should give us hope, they can't be cause for complacency. We're doing more to stabilize our planet than most people thought possible just a few years ago. And at the same time, the honest truth is that we're still not doing nearly enough.

While a growing number of brilliant scientists, engineers, businesspeople, activists, and political leaders are devoting themselves to protecting our planet, I would never want you to think we're going to win on climate because of people who are smarter than you. We're going to win on climate *because of you.*

Perhaps you read that last sentence and wondered if that's really true. Can you personally make a meaningful difference? Trust me. You can. In fact, now is the perfect time to change your life in a way that reflects the realities of our changing world. Stabilizing the planet is the fight of our lifetimes—and it's a fight we need you to join.

How, specifically, can you have the greatest impact? I think it comes down to three areas: personal choices, active citizenship, and purpose-driven careers.

By "choices," I mean the individual actions we take as consumers. There are a lot of small ways you can help reduce humanity's overall emissions. In its document "Actions for a Healthy Planet," the United Nations suggests a dozen ways you can help the world meet the Sustainable Development Goals laid out in 2015. They include everything from buying an electric vehicle to installing an electric heat pump in your house to throwing away less food.

These are all good ideas. In many cases, despite the reams of fossil fuel propaganda trying to convince you that doing your part on climate will be expensive, making climate-conscious choices will both save you money and improve your quality of life. Also, the choices you make as an individual do have impacts on overall markets, even in small ways. If you're an early adopter of cleantech, you're helping start-ups grow more quickly. If your neighbors see you driving an electric car, putting solar panels on the roof, or saving money with a heat pump, they're more likely to do the same. For all these reasons, if you want to become a climate person, an easy first step is to ask yourself, "How can I, personally, get closer to net-zero emissions?"

But as I've said, while reducing your carbon footprint can be the beginning of the change you make to your life, it can't be the end. I worry that it's naïve, and perhaps even counterproductive, to suggest that solving climate starts at home—it makes it seem like climate is purely a matter of personal responsibility, shifting blame to consumers and letting the fossil fuel industry off the hook.

Focusing on individual choices would be like saying, thirty years ago, "Lung cancer is a worldwide public health crisis, so you need to quit smoking." Quitting smoking is a good thing to do. But it's not a solution to a global problem, and it's no match for corporations who make billions of dollars perpetuating that problem for as long as possible.

That's where citizenship comes in. I'm not referring to the kind of citizenship that comes with a passport. Instead, I use citizenship in the sense of active, engaged participation in your community. Citizenship is what drives collective action, especially in democracies. Which means that harnessing your power as a citizen is one of the most effective things you can do to help stabilize our planet.

Citizenship takes many forms. Among the most direct is to get involved in politics. Party politics has a bad reputation these days— and believe me, I can see why. But as the saying goes, elections have consequences. Consider the most recent presidential race. Joe Biden wasn't my first choice for his party's nomination, in no small part because I was running against him. As president, he's made some climate-related decisions, such as approving the Willow oil project, that I definitely wouldn't have made.

But it's hard to overstate the impact that American voters had when they elected Joe Biden in 2020—and could have again this year. A bipartisan infrastructure bill that helps modernize everything from mass transit to our power grid. The Inflation Reduction Act, the single biggest investment in cleantech in world history. Denying fossil fuel companies the right to drill on millions of acres in the Arctic, including on public land leased for drilling by the Trump Administration. All this, and so much more, would never have happened if the outcome of the last presidential election had been different. President Biden and his administration have been the strongest proponents of a robust climate response in history. By far.

And in this coming election, the difference between two presidential candidates on climate has never more clear. On one hand, you have a president, and a party, determined to build on the successes of a first term—bringing down energy costs, supporting new technologies, and helping get them into the hands of consumers nationwide. On the other, you have a candidate just ready and waiting to put Project 2025 into practice, throttling clean energy, and putting the full weight of the federal government into propping up oil and gas. Donald Trump was the worst climate president in history. He's promised, proudly, to be even worse if he gets another shot—and the Republican Party is determined to help keep that promise even if Trump himself doesn't wind up back in office.

Today, you don't have to love politics to get involved in politics. You just need to care about preserving a livable planet. We can't afford a four-year-long pause in the clean energy revolution, let alone a giant, fossil fuel–funded step backward. Whether it's encouraging friends and family to vote, knocking on doors, donating money, using your social media (including the original social media—having conversations) to counter misinformation, or amplifying climate news that helps people understand what's at stake, there are so many ways to be an active citizen in an election that, in many ways, could be the fossil fuel companies' last stand.

Also, while presidential politics gets most of the attention, that's far from the only place where citizenship makes an enormous difference. In some ways, oil and gas has even more influence over statehouses than they do in Congress, because they can hire lobbyists to pursue their interests with less public scrutiny than they'd receive at the federal level. You can help change that. Help cast a light on what's happening in state government by writing an opinion piece or a letter to the editor. Get involved with a local campaign. Or even run for office yourself.

Nor is citizenship limited to electoral politics. Take my home state of California. These days, we're a reliably blue state, but putting into place new laws and programs to help stabilize the planet still requires political will. Just last year, our legislature was debating a bill that would require businesses operating in California to disclose its Scope 3 emissions—emissions from every piece of the supply chain. Because just about any company that wants to operate at scale has to do business in California, passing this bill represented a huge potential win for climate measurement not just in the Golden State but around the world.

As usual, the oil and gas industry fought against this initiative with everything they had. What made the difference was the grassroots movement on the other side. Ordinary people, some acting on their own, some through organizations, encouraged local lawmakers to go big. They made sure assembly members and state senators knew that they'd be held accountable if they sided with fossil fuels over their constituents, and that if they did the right thing, a grassroots army would have their back. And while the fossil fuel industry and its allies like to pretend that climate is an issue that only rich, overeducated white yuppies care about, the polling showed that in California—just like in the rest of the country—African Americans and Latinos care more about taking action on climate than any other group.

At the same time, some of the biggest companies in the world stepped up to provide a business counterweight to oil and gas. They understood that being on the right side of this issue isn't just good for the planet or for future generations—it's hugely important to their customers and employees. Together, California demonstrated how change happens. These kinds of efforts are taking place around the country and the world. No matter where you live, they're something you can be part of.

Finally, citizenship frequently doesn't involve politics and government at all. While small-scale initiatives within a community don't have the same influence as nationwide or statewide laws, often an effort that begins in just one neighborhood can be the start of something that changes the entire country, or even the world. Whether it's protesting or organizing in person, or using your platform on social media, there are always opportunities for climate people to be engaged citizens.

At the same time, however, most of us don't have the luxury of being full-time active citizens. We have jobs. And this brings me to the final, and perhaps most important, area where you can make a difference on climate: your career.

Because our work takes up so much of our time, and, for many of us, makes up so much of our identities, career is the element of becoming a climate person that requires the most commitment—and that many people are the most resistant to embrace. I get it. It took me seven years to go from being concerned about climate to quitting my job and focusing on climate full-time. Looking back, though, my only regret is that I didn't make the change sooner. Building a business was challenging, exciting, and fun. Success for success's sake feels great. But it's nothing compared to the feeling that comes with having a meaningful career, knowing that each morning brings another opportunity to be a small part of solving one of the biggest problems in the history of humankind.

In short, choosing to pursue a purpose-driven career was the best professional choice I ever made. And you should make it, too.

For young people, figuring out how to work in climate is comparatively easy. There's a saying I've sometimes heard among comedians when they talk about going into comedy: "If you can be happy doing anything else, do that." For climate, I'd turn this on its head. If there's any way you can see yourself being happy while working on climate full-time, you should. That's not going to include everybody. If you know in your heart that you should be a concert pianist or a videogame designer, I'm not here to tell you to live a life you won't find fulfilling. But if one of your potential professional options involves helping to stabilize the planet and protect humanity from the worst global crisis we've ever experienced, and the others don't, it seems obvious which career path you should take.

What's more, building a climate-focused career has never been easier. It's impossible to say exactly how you should incorporate climate in your work, because each of us has a different set of talents, experiences, and skills. In most cases—whether you're just starting out or making a pivot—it's possible to do what you love, and what you do well, as a full-time member of the climate movement. If you're a marketing executive, work for a company in cleantech. If you're a teacher, build climate into your lesson plans and make sure your students understand the opportunities in cleantech as they consider their future careers. If you're in sales, help sell products that reduce carbon pollution. (And to go back to the example from the last paragraph, if you're a successful concert pianist or videogame designer, use the platform you'll develop to educate people and tell stories that help drive collective action on climate change.)

Back when I first became involved full-time in climate, helping to stabilize the planet often meant making less money. That's not true anymore. Today, in most cases, making the change to a purpose-driven career won't just be good for your life. It will be good for your bank account, too. Here are just two examples, from very different professions. If you install insulation, and you decide to specialize in retrofitting buildings so that they require less oil and gas to heat and cool, you'll be one of the earliest people to tap into a huge, growing market. If you're an attorney, and you decide to help hold the fossil fuel industry accountable for the harm it's done to the planet, you'll be setting yourself up to receive a portion of what will likely be the largest financial settlement or judgment in history.

I'm not saying that working on climate is always lucrative—nor should it be. But if you want to do well by doing good, this is your moment.

That's even true—and perhaps especially true—for people who work in oil and gas. The more time I spend on climate, the clearer it becomes to me that the fossil fuel industry will never willingly take responsibility for the damage it's caused. We'll have to beat them: in the markets, in the marketplace of ideas, and at the ballot box. But in my experience, once you beat people, they tend to join you. In fact, as far as I'm concerned, the ultimate sign of victory isn't when your opponent says, "I give up." It's when they say, "I was on your side all along."

It may seem hard to imagine right now, but we will reach that moment—and if enough of us make big changes to our lives, we'll reach it sooner than most people think. I'll never be afraid to tell the truth about what the fossil fuel companies are doing to our country and our planet, or to criticize the people who somehow convince themselves that working in the oil and gas industry is worth it. But already, we're seeing more and more people take the knowledge and connections they acquired working in fossil fuels and put them to work in cleantech. Sometimes they're motivated by a sense of right and wrong. Other times, they just want to leave a stagnant industry for a rapidly growing one.

Regardless, I hope we welcome them with open arms. Because while climate is the fight of our lifetimes, we are all ultimately on the same side. Stabilizing the planet—and making energy cheaper and more reliable in the process—is good for everyone.

Throughout this book, I've talked about getting to net-zero emissions. That must be our top priority. Until we achieve it, every other piece of progress rests on shaky ground. But what *really* excites me is what happens next. Because once we've stabilized the planet, together we'll be able to build the kind of future our parents and grandparents could only dream of.

The clean-energy revolution will, of course, have a role to play. Once we reach net zero, it will no longer be difficult to go even further, taking more carbon out of the atmosphere than we put in. I don't want to sugarcoat it: some of the natural systems we've disrupted can't be completely repaired, and because greenhouse gas remains in the atmosphere for so long, extreme weather will continue to grow worse even after we've hit net zero. We're not going back to the climate of 1850, or even 1950. That's just a fact. We'll still have to mitigate the harm from climate change.

But once we've gotten to net zero, and start getting net negative, at least we'll know what to expect, and how to mitigate it. And many of the worst effects of a warming planet *are* reversable. It won't happen overnight, but we can remove some of the most dangerous threats humanity faces.

At the same time, the clean, cheap, reliable energy breakthroughs required to get to net zero will completely revolutionize our quality of life. Imagine if households could save nearly all the money that goes toward electricity and heat; if manufacturing could be done for a fraction of the cost. So many of the problems, large and small, which today we accept as facts of life are in fact a result of energy being expensive and coming from pockets of long-dead plants and animals buried randomly under the ground. Clean, cheap, widely available energy isn't the stuff of science fiction anymore. We can make it real. The economic impacts of the energy revolution are going to be huge.

But human beings aren't just widgets in an economy. Just as so much of what makes life worth living can't be tallied on a spreadsheet, some of the biggest effects of the transition to clean energy won't be measured in dollars and cents. Just think about how many wars have been fought over oil, and how often oil-producing nations are able to hold the rest of the world hostage to their demands. Now imagine a future where every country on earth can produce its own energy, in nearly unlimited quantities—think of all the conflicts that we'll be able to avoid. Think about how that levels the playing field for people all over the world. Think about how it changes global power dynamics.

Then there's health. Burning oil and gas doesn't just send oil and gas into our atmosphere. It also spews out all kinds of other compounds that are unsafe to breathe, leading to everything from asthma to cancer to lung and heart disease. According to the World Health Organization, air pollution claims 6.7 million lives per year, and other studies have put that number even higher.

The fossil fuel companies like to talk about cleaner gas, or cleaner oil—just like they used to talk about clean coal—but that's an oxymoron. They're doing the same thing the cigarette companies did when they tried to get people to switch to "filtered" cigarettes instead of quitting smoking. There's just no way to get around the fact that their products are killing people, especially in some of our poorest and most vulnerable communities—and when the clean energy revolution is over, that's finally going to end. We won't just live on a safer planet. We'll be breathing cleaner air and drinking cleaner water.

There's also the broader question of our relationship to the natural world. The fossil fuel companies don't just represent an industry. They represent a mindset focused on extraction—that the only way to enjoy the benefits of the modern world is to destroy the natural one. This idea, that plundering is an unavoidable part of life, is a recipe for tragedy and disaster. It's also clearly unsustainable. The clean energy revolution represents a completely different mindset, one that says that we can have more while taking less. I know which planet I'd rather live on—and which planet I'd rather hand down to my children and grandchildren.

Finally, there's the simple but crucial fact that once we've stabilized the planet, climate change will no longer jeopardize all the incredible progress we've made in other areas. In my lifetime, and even in my children's lifetimes, we've seen enormous breakthroughs in medicine, treating diseases from cancer to Alzheimer's with success that would have been hard to imagine not so long ago. Worldwide, the percentage of people living in extreme poverty has plummeted, while the number of people with access to education has skyrocketed. Stabilizing the planet gives us the chance not just to continue those trends but to accelerate them.

It can be hard, in the midst of a crisis, to think we might not just survive the moment but emerge from it on the cusp of something remarkable. But that's exactly what we've done before. At the start of this book, I talked about how members of my father's generation asked one another, "What did you do in the war?" For each of them, the answer was different. But there's another way to look at that question. Collectively, my parents' generation beat back the greatest threat the world had ever known and gave their children and grandchildren the chance to live amazing, interesting, extraordinary lives in the greatest country on earth.

If we come together in this moment, then one day our grandchildren will be able to say the same thing about us.

ACKNOWLEDGMENTS

There were many people who contributed to the writing, editing, and publishing of this book.

First and foremost, David Litt did yeoman's work in turning my thoughts into polished sentences, paragraphs, and chapters. Plus he's funny.

Cindy Spiegel and her team at Spiegel & Grau have been guides and wise counselors all along the way.

Bob Barnett has been a properly aggressive and very wise representative. Nothing would have happened without him.

The team of people in San Francisco includes a host of dedicated and smart people, including my long-time compañeros Alex Fujinaka, Zack Davis, and Aaron Burgess.

I could also never leave out two of my closest friends and coworkers, Shannon Hart and Brooke Rice. They help and protect more than anyone deserves.

And of course, I want to thank my partners and coworkers at Galvanize, led by my first professional partner and close friend, Katie Hall. I play on an amazing team, both in terms of professional competence and climate integrity. We're supported by strategic investors with the highest values and brains.

And then, there are literally hundreds of people in the community responding to the climate emergency—NGO workers, teachers, activists, investors, entrepreneurs, and policymakers. You have inspired, taught, comforted, befriended, and partnered with me. This community makes the work fun and exciting. Let's prosper and succeed together, and may our community only grow.

ABOUT THE AUTHOR

TOM STEYER is the founder and coexecutive chair of Galvanize Climate Solutions, a multistrategy investment firm focused exclusively on the energy transition. After earning his MBA from Stanford, in 1986, Steyer went on to found Farallon Capital Management, a San Francisco–based hedge fund that pioneered the strategy known as "absolute return investing" and that grew to $20 billion in assets under management at the time he left. In 2012, he left his firm to devote his time, money, and energy to climate issues. Steyer played a key role in preserving California's Global Warming Solutions Act while also working to pass clean energy initiatives and advocating for environmental justice across the country. With his partner, Kat Taylor, he cofounded Beneficial State Bank, a triple bottom line community development bank focused on justice and sustainability, and TomKat Ranch, a regenerative ranch dedicated to raising cattle with a negative carbon footprint.

In 2013, Steyer founded NextGen America (formerly known as NextGen Climate), the largest youth voter engagement organization in American history, whose climate-focused messaging and outreach helped lead to record levels of youth turnout in recent elections. He was a 2020 Democratic presidential candidate with a campaign centered on addressing climate change, and later that year he served as cochair for California governor Gavin Newsom's Business and Jobs Recovery Task Force. In addition, he cochaired President Joe Biden's Climate Engagement Advisory Council to help mobilize climate voters.

He lives in San Francisco, enjoys spending time with family, and can always be counted on by friends to relay the latest climate data (whether they are interested in hearing it or not).